ICELAND BY KAYAK

ICELAND BY KAYAK

THE FIRST CIRCUMNAVIGATION OF
ICELAND BY KAYAK

Nigel Foster

NIGEL
KAYAKS

© nigel foster 2023
All rights reserved.
ISBN-978-1-7364203-2-4 (paperback)
ISBN-978-1-7364203-3-1 (hardcover)

Front Cover photo: Axel, right, talks with Geoff at Gjögur, northwest Iceland.
Back cover photos: Fishing boat, Aron, leaves Flatey Island.
Nigel Foster on Flatey Island, 1977, by Geoff Hunter.
Cover photographs and all images in this book by Nigel Foster and Geoff Hunter, except where otherwise credited.
Maps by Nigel Foster.

Publisher Nigel Kayaks
www.nigelkayaks.com

To Geoff Hunter

Contents

Nigel Foster iii
Acknowledgments ix
Foreword by Geoff Hunter xi

Introduction xv
1   Wormholes 1
2   Varnish and Canvas 7
3   Sea Kayaks 15
4   New Kayak for New Plans 25
5   Tie the Paddles 39
6   Faeroe Islands 53
7   Seyðisfjörður 61
8   A Taste of Wind 77
9   Pinned 87
10 Höfn 101
11 Ingólfshöfði 111
12 Into the Sands 125
13 The Long Wade 137
14 Vík 147
15 Breezy Again 155
16 Vestmannaeyjar 167
17 Salmon 179
18 Reykjanes 189
19 Keflavík and Reykjavík 201
20 Faxaflói 213
21 Breiðafjörður and Beyond 229
22 Suðureyri 241
23 Rounding Horn 249
24   Gjögur 267
25 Siglufjörður 279

26 Flatey Island 291
27 Fox Plains 305
28 Langanes 325
29 East Coast 339
30 Seyðisfjörður Again 347
    Epilogue 351
    Background reading 353
    More titles by Nigel Foster 354
End Notes 357

Maps

1 Iceland .... xiv-xv
2 The Route North from Burwash to Seyðisfjörður 41
3 Faeroe Islands 59
4 East Coast of Iceland 60
5 South Coast of Iceland 100
6 Southwest Iceland 188
7 Northwest Iceland 228
8 North Iceland 266
9 Iceland, Siglufjörður to Langanes 290
10 Northeast Iceland 324

Acknowledgments

First, my thanks to Geoff Hunter, a wonderful person with whom to share an adventure, and to your wife Jeannie who coordinated from England during our trip. Also, Valerie Harrison for all your support, including making drybags for us on an old sewing machine in my tiny attic room.

Throughout the planning process and the trip, others helped us achieve our goals by offering equipment, or advice, or by extending friendship and hospitality. That includes our sponsors, such as Lendal, Damart, Tupperware, Wild Water, and Bell and Howell, and individuals who appear in the story. Named here or not, we very much appreciated your help. I particularly thank those Icelanders who made our journey so special. To Ken and Sheilah Gulliver of Burwash Place, my deep gratitude for the faith you showed in me.

In completing this book, I relied heavily on my journals. My notes were the key to organizing myriad memories that must surely have otherwise become jumbled. I have my parents Peter and Elizabeth to thank for tirelessly coaxing me to keep a diary, encouraging me to set a goal of a little each day. From little steps come long journeys.

Michael Muller, thank you for mentoring and especially for pointing out ways to better use Microsoft Word.

Finally, Kristin Nelson, my wife, thank you for proof-reading. Your encouragement, and patience, while my mind dwelled in a far-off country, in the past, as much as in your time and space, was precious. You are the best!

Foreword by Geoff Hunter.

Nigel contacted me, in late 1976, when he had the vision of circumnavigating Iceland by kayak. He had played a huge part in designing and building the kayak that we would each use: the Vyneck. It was fast, and quite tippy when empty. The keel was straight and low with little rocker, so it was not easy to turn. Roomy enough to carry a lot of cargo, it proved to be more stable when loaded. This was perfect for us on our trip as we were going to some remote places and needed to carry not only our gear for camping, but food and provisions for a week or two without passing any habitation.

We hadn't paddled the kayaks much, or even paddled together before we left for Iceland, but it didn't take us long to settle down with the boats and with each other.

Iceland has a geography of such varied coastline, from the flat, dangerous, and inhospitable south coast to the huge cliffs and deep fiords in the West and North. Landing proved not always easy or even possible. For a sea kayaker, it was an incredibly exciting and beautiful country to explore.

Nigel was the most capable and skillful canoeist that I had met and a perfect partner to paddle with on an expedition to a place where sea conditions are so demanding. We both became very comfortable controlling the boats in some very difficult, tough, sea states and seemed to be good companions in working through the unexpected challenges that the landscape presented. This was especially evident on the desolate and isolated mudflats and lagoons on the south coast.

Throughout our trip, we found Icelandic people to be welcoming and generous. Often invited to visit or stay with them, we never went hungry.

Nigel writes in a very descriptive and informative way, making it easy for me to recall what a wonderful journey we shared together.

I shall always remember that summer with great fondness, and I feel honored to have been invited to have taken part in such an epic and exciting adventure. Thank you, Nigel, for having the vision of the trip in the first place and for bringing all the memories back with your excellent book.

Geoff Hunter.
2023

GEOFF HUNTER IN 1977.

Introduction

Ask your smart phone and it can list the kayak rental options in another country. With the same ease you can discover the cheapest flights, ferry fares, others' trip reports, and even where to get coffee at your destination. You can summon Google Earth for satellite images of a coast you would like to explore and spot places to camp.

At a finger touch you can download or print sea charts, information about tides and currents, swell reports, and world weather. Shopping online, you can select kayaks and paddles, drybags and dry-suits, tents, camping accessories, and more. You can translate between languages. Nowadays you might plan a whole expedition while relaxing at your local café. On the water you can use GPS (Global Satellite Positioning) to see where you are, even in fog. You can maintain contact with others using a waterproof VHF radio, cellphone, or satellite phone, and obtain updated weather forecasts. Yet it was not always so easy.

The Vietnam War ended in the year *Saturday Night Live* first aired, and Bill Gates founded Microsoft with Paul Allen. The following year, 1976, Steve Jobs and Steve Wozniak incorporated Apple Computer. Home computers were not yet on the market.

Until 1976, every telephone in the UK had a circular disk with fingerholes numbered zero to nine. A thick coiled lead tethered the handset to the weighty body of the phone, cabled to a wall socket. On lifting the hefty handset from its cradle to hear the dialing tone, you inserted a finger to rotate the dial, removing your finger after each number in the sequence to let the spring-loaded disk spin itself, purring, back to rest. Commonly, the doorstop of a local phone directory lurked near the phone.

Then, the new *key-phone* came to the UK. It had a rectangular keyboard of buttons on the phone's base to achieve the same result as the dial. There were no mobile handsets.

In 1976, long before the World Wide Web, I began planning to kayak around Iceland. I had to ask a high street travel agent how to get there. The local public library was my next indispensable source of information with its drawers of typed cards, referencing the books and journals where I might find something pertinent. First-hand advice was thin and often irrelevant.

Comparing Icelandic topographical maps with sea charts, I saw how often the spelling of place names varied. This stemmed from the difference between English and Icelandic alphabets. Phonetically transcribing, using a different alphabet, results in multiple interpretations. So, which spellings should I use in this book?

My wife's name is Kristin, and I know how important it is to differentiate between Kristin and *Kristinn, Kristen, Kirsten, Kerstin, Christine, and Kristine*, all sounding similar. So, out of respect, consistency, and I hope accuracy, I have used the thirty-three letters of the Icelandic alphabet for Icelandic names and places wherever possible in this book. Incidentally, the letters c, q, and w, although not in the Icelandic alphabet, appear in foreign words commonly used in Iceland. Placename spellings are from the Icelandic maps we used in 1977, which may differ from those shown on British charts, American atlases, and web sites.

Icelandic spelling is phonetic. In English there is no difference between how you write the *th* sound in the words *the,* and *thistle,* although they sound different. When we say *the,* we voice the th sound, but when we say *thistle,* we do not. Icelandic uses Ð or ð for the voiced sound and Þ or þ for the unvoiced.

British Admiralty charts change ð to dh, as from Seyðisfjörður to Seydhisfjördhur, and Þ to Th, as from Þorlákshöfn to Thorlákshöfn to approximate the pronunciation. Conversely these charts still display the letters á æ é í ó ö ú ý.

English language atlases and web sites usually omit all letters outside the English alphabet, making spellings visually approximate but less phonetically correct. In this way Seyðisfjörður appeared as Seydhisfjördhur on our British charts and Seydisfjordur elsewhere. If you choose to follow our progress using a large-scale map, you will see what I mean. Such spelling inconsistencies aside, as a Brit living in the United States I write in a mix of British and US English. Nothing is ever simple.

I have tried to portray things as they were in the 1970s. Tourism is just one example of something that has changed since. In 1977 there were 72,690 visitors to Iceland, almost one-third the 222,000 population. In the next ten years the annual number of visitors reached 129,315, increasing by 1997 to 200,000.

By 2017, with the number of Icelandic residents at 333,000, the annual influx of visitors had soared to more than a million, more than three times the population. Places, infrastructure, economy, attitudes, and empty spaces all changed accordingly.

In 1977, we saw just one other kayak in Iceland. Now there are thriving kayak clubs and rentals. Where boats left to hunt whales in the 1970s, tourists now join whale-watching tours. Cars circle the country on asphalt, instead of on a washboard of potholed graded gravel. Roads contour around headlands inaccessible twenty years ago. Ferries from a new south coast ferry port make a short crossing to Vestmannaeyjar, *Westman Islands*. Compare the descriptions in this book against a modern map, and against today's data, and you can spot the differences.

Here is an adventure: the first circumnavigation of Iceland by kayak, in 1977, and how it came about.

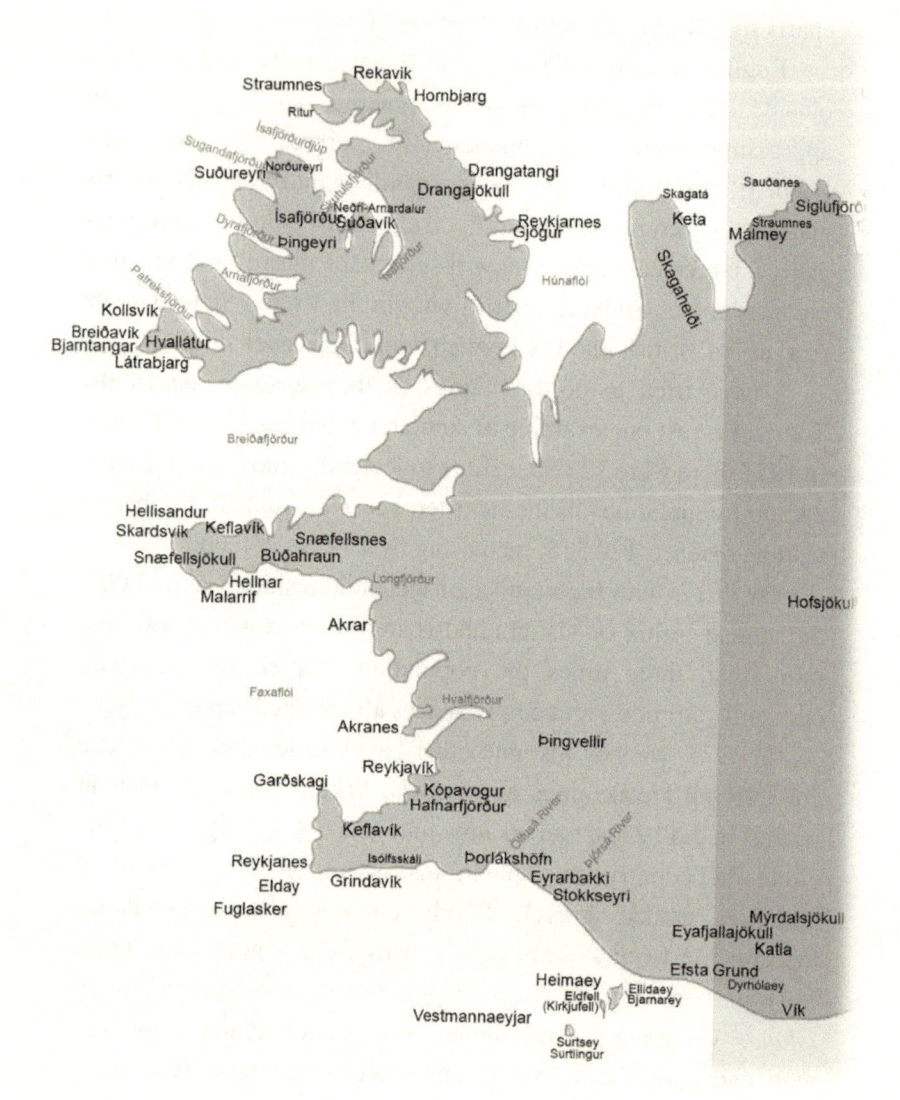

MAP 1. ICELAND.

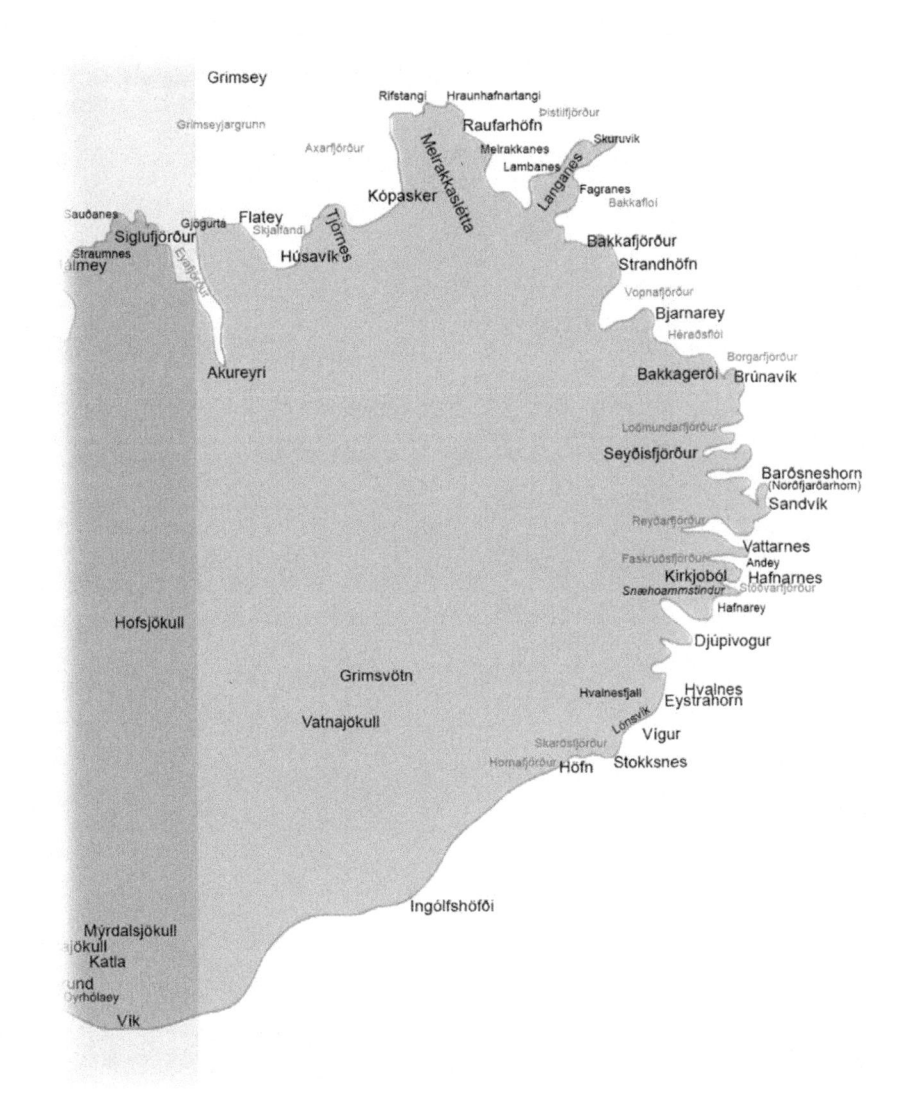

1
Wormholes

I took an extra deep breath. It was my turn now. "Take off your helmet and turn out your light." I meekly obeyed. As the hissing acetylene flame of my headlamp dimmed to a glimmer, popped, and extinguished, my shadow, cast by the lamps behind me, shuddered into life against the cave wall. I dreaded what would come next.

Kneeling on the clay floor as if to pray, I pushed my helmet ahead into the dark hole and wriggled head-first after. "Stretch your arms straight," came muffled instructions from behind, my body in the entrance blocking the sound. "Don't use your elbows or you'll get stuck, just use your feet." The darkness of the tunnel closed around me as I wriggled forward.

Now the sound of my breathing filled the tiny space. My clothing rasped against the hard mud; my helmet scraped as I pushed it ahead. My hair dragged along the roof. Abruptly my shoulders jammed against the sides. Breathing out, I lowered my cheek to the floor and stretched myself thin. As my chest narrowed

Iceland by Kayak

and my elbows straightened, my shoulders released from the sides of the tunnel.

I wriggled forward again, pushing with the toes of my boots against the clay. I tried to control my breathing without fully inflating my lungs, tried to keep from panicking in the dark. Instinctively hauling forward with my elbows, my shoulders jammed. I gasped in panic, stuck. I reasoned with myself and eased backward to release. Forty feet, they said. A wormhole forty feet long.

My neck already ached. I laid my head down onto the mud for a moment to ease the muscles, and then lifted it and pushed forward with my toes, rocking my body until I felt myself move. I had to force myself to keep my arms straight out ahead and not bend my elbows. I wriggled inch by inch, pushing my helmet ahead. It scraped hollowly.

Impatient to move faster, I could not. I must stretch, wriggle, push with my toes. True, I was moving, just barely. Forty feet. I was on the verge of panic, reasoning against my instincts. If the others had made it to the end, I could too. I could taste my breath against the damp mud, feel the moisture hanging in the stale air.

I wished for light. Any light. I thought about all the weight of rock above me. This muddy hole was narrower than my normal shoulder width, so I could pass through only if I held myself thin, although that made it difficult to move forward. My heart raced and I tried to relax. I had to believe I could make it, for I could imagine the consequence of panic. All at once I could hear muffled voices. Surely, I had not far to go.

At last, I felt my helmet lift from my fingers, heard a low chuckle. "You're almost here." A hand touched my head. I lifted my face and thumped the roof. "Keep low or you'll get stuck. Come on, you're almost here."

As I slithered out, hands guided me in the total dark. The air smelled stale and moist. "Okay, sit up, turn around, and come

Wormholes

back. There's a wall." Someone lit a small flashlight revealing three figures, hunched against the walls, their breath misting into the tiny cavern that looked hollowed out of clay. My eyes snatched thirstily at the steaming mud-caked boiler suits and wrinkled leather hiking boots. Metal lamp reflectors gleamed from dark helmets above pale faces that shadowed into slumped forms, ethereal shapes jumping as the flashlight jerked. There was no way onward. A dead end. The light vanished abruptly, leaving bright patterns in my eyes. "We can't use the carbide lamps, or we'll burn up all the air. I'm saving my battery in case."

In case of what, I wondered? A dead end. We must go out the same way. I should never have come in. There was an awkward chuckle, "Let's get the others through so we can start moving back out before we run out of air."

He shouted down the tunnel, "All clear! We're ready for the next!" The answering call came back flat, distant, deadened by the damp clay walls. Somebody was starting through.

Caving, or potholing, was both love and hate for me. I loved the wonderland of stalactites, and stalagmites, and the smooth flow of calcite spread like ice. I loved the colors in the rock, and the vastness of the larger chambers, too deep and wide for a lamp to fully reveal. But I never really enjoyed crawling through cold mud, getting drenched, scraped, and stinky, wriggling through tight spaces that threatened to snare me.

I forever dreaded facing a sump, a u-tube of tunnel filled with water. A sump would require me to extinguish my lamp and follow the roof underwater, to surface at the far end, hopefully before I ran out of breath. Scared to put my face underwater at the best of times, even the thought of a sump, in the dark, sent my heart racing.

After each caving trip, cold on our journey homeward, I sat with the others on inward facing bench seats, huddled from the night wind that buffeted the canvas back door of the soft-top Land

Rover. Swaying and leaning as the Land Rover negotiated the winding roads, tote bags, sacks of wet boiler suits, and fresh-rinsed coiled ladders held us in place.

As we sped through the lanes, I always swore never to go again. Yet, paradoxically, I would sign up for the next caving trip, to inexorably face my doom like a passenger on an escalator regretting that first step until released at the top. I began to realize how, drawn to confront my fears, to banish or come to terms with them, I ratcheted up the level of challenge and discomfort step by step. I was stretching the bubble, extending the boundaries.

My parents encouraged us children to try various activities. This despite an occasional surprise such as learning that a bat I had captured, and secretly brought home, had escaped into our basement when I opened its box.

After crawling underground, how about rock climbing? As a child I feared heights. I began climbing on a low cliff, in the south of England. The sandstone outcrop was the result of a geological fault. For safety, we looped a hawser-laid hemp rope around a tree at the top and flung both ends down the face.

With one end tied firmly around my waist with a bowline knot, I climbed. My climbing partner, looping the other end of the rope behind his waist, used both hands to reel in, keeping the rope taut while I climbed. Should I slip, I expected him to cling on tightly to prevent my fall; a promising idea that did not always work. Hemp has little stretch and is abrasive to the hands. The jolt of a falling climber, with the sudden burn of hairy rope dragging through tender palms, is enough to startle any neophyte into letting go of the rope.

I considered the daylight and open air of rock climbing more appealing than the confinement of potholing. To protect the trees, I began using rope slings with steel karabiners, before upgrading to lightweight aluminum karabiners and nylon webbing slings. I graduated to nylon kernmantle climbing rope which was easier on

Wormholes

the hands. It was also less brutal to the climber than hemp, easing the jolt with a little stretch in the event of a fall, something especially welcome in those days before the climbing harness. I climbed now with equally motivated partners.

Finding I could climb sandstone better with bare feet, I began to enjoy the thrill of free climbing the easier routes: that is climbing without a rope, until scary ascents on wet rock helped dissuade me. But bare feet and top roping became impractical when I began tackling taller cliffs in the Welsh mountains and on the Cornish coast. There, with climbing boots and safety hardware, I learned how to lead multi-pitch climbs.

By this time, I had evaded puncture while learning to fence, tried sweep rowing, and swum from overturned rowing eights. And then, despite or because of being an unenthusiastic swimmer, I became hooked on kayaking.

CARBIDE CAVING LAMP.

Iceland by Kayak

AUTHOR, LEFT, WITH HIS BROTHER, MICHAEL. BRIGHTON 1968.
PHOTO BY PETER FOSTER.

2
Varnish and Canvas

My very first kayaking experience was in a canvas tandem kayak, on an oxbow lake, near the mouth of the River Cuckmere in Sussex, England. The Cuckmere River, with its source in the rutted sandstone and clay landscape of the Sussex Weald, wends its passage south through the chalk hills of the South Downs.

The oxbow was once a meandering stretch of river, little more than a mile long, at the edge of the Downs. It has become a shallow cut-off, while the straightened section of tidal river hurries under the coastal road, briskly by-passing this lazy backwater along a cut directly to the English Channel.

I was fourteen years old, following my brother Michael who had already tried kayaking. I cannot recall how the leviathan of a kayak, together with an equally hefty single kayak, came to this location. I expect the Land Rover which ferried us to the Mendip caves also brought the kayaks. Michael, a seasoned paddler in my eyes, paddled away confidently in the solo kayak as I stepped gingerly from the warm, shallow water into the second seat of the tandem. Framing the inside of the kayak was a deeply varnished wooden skeleton, darkened in places where water had soaked

Iceland by Kayak

through cracks in the varnish into the wood beneath. Covering this frame, and tacked to it around the gunwale, was canvas. A generous coat of paint sealed the canvas. There was a distinctive smell in the steamy-warm air, of stale damp wood, canvas, and varnish.

I lifted the long wooden paddle and followed instructions as we drifted gently from shore. Beneath me the water writhed and wriggled with pond insects and bubbling waterweeds. Dragonflies darted above the water, chasing mosquitoes and gnats. We did not paddle far, still the experience delighted me so much I could not wait to try again.

The Scouts offered my next opportunities to paddle, on the rivers Adur and Arun, touring in canvas kayaks, and camping, but my commitment cemented when I bought my own kayak. This was a used, skin-on-frame, *PBK-11*, a model originally sold as paper plans or kit by designer Percy Blandford. It was eleven feet long, hence PBK-11, *Percy Blandford Kayak 11-feet-long,* and had a plastic-coated canvas skin, lightweight white deck, and tougher grey hull, tacked over a skeleton of varnished plywood. Michael had sparked me into action by buying an unfinished PBK from a neighbor, for the grand sum of one pound. Following up a classified ad in the local newspaper, I bought mine intact, and at the same time inadvertently gained a paddling friend, the seller, Jez. But a kayak is of little use without access to water, and our family home was two-and-a-half miles inland, at 360 feet elevation.

Nothing deterred Michael. "We'll make trolleys," he announced enthusiastically. "Then, we can tow the kayaks behind our bicycles." Taking apart an old pushchair, a small-wheeled children's carriage, he used the wheels to make a kayak cart. Soon we were both mobile, towing our kayaks down the valley to the closest seaside village, historic Rottingdean, where the lane narrowed and twisted, slowing us to a crawl.

Varnish and Canvas

When Rottingdean appeared in the Domesday Book, in 1086, twenty years after the Norman conquest, it had already existed as a Saxon settlement for at least five hundred years. A narrow gap in the chalk cliffs offered access to the sea. It was through this gap that the French attacked in 1377, to lay waste to the village and beyond. Before leaving, they set fire to everything, including St Margaret's Church. All those villagers, who had locked themselves inside the stone tower for safety, burned to death. The church still bears the scars in the heat-discolored flints.

Across the lane from the church, before we reached the village duckpond, stood *The Elms*, a property once rented by the writer Rudyard Kipling. When Kipling left Rottingdean to escape the admiring but nosey crowds, he set up home on the quiet outskirts of another East Sussex village, Burwash, a village that was later my own home for a time.

Finally, navigating through the awkwardly tight and crowded High Street which never adequately accommodated motorized traffic, we negotiated our way to the shore. There, in the damp salty breeze that always funneled through the gap, we leaned our bikes against the sea wall beneath the white chalk cliff. Securing our cycling clothes in large, sturdy, polythene fertilizer sacks from a nearby farm, rinsed clean, we closed the bags with string, folding the necks double and tying again for good measure. With the bags stuffed into our kayaks for safekeeping, we donned lifejackets and launched from the flint cobble beach.

There was something wonderful about the motion of the kayak through the water, its smooth glide. The gentle flex of the hull over small swells felt like the breathing of something alive. I cherished the freedom to meander close along the coast beneath the chalk cliffs, or to venture far from shore.

After each excursion we toiled back up the hill. I could never keep up with Michael, and trudged behind, pushing both bike and kayak. When Michael progressed to other projects, I made the

Iceland by Kayak

easy decision to dispense with the bike and simply walk the kayak each way.

In my independence, I dallied offshore, wondering what went on inside Roedean College, the residential public girls' school that stood so regally atop the cliff. Nearby Black Rock was another place that sparked my imagination. There, a 200,000-year-old raised beach deposit filled a former valley in the older Cretaceous chalk. The deposit held the remains of mammoths, woolly rhinoceros, and tools from Neanderthals. Usually turning around on reaching Brighton's Palace Pier, sometimes the extra mile of crowded holiday beaches to West Pier tempted me onwards, making my round trip about eight miles.

At school, after summer break, taking every opportunity to attend courses at a residential outdoor activity center, I sought ways to specialize in kayaking in preference to climbing or hiking. In those days, the center kept a fleet of sleek touring kayaks, with fiberglass hulls and plywood decks, but kayaking trends were changing. My first taste of kayak-surfing soon followed, in a fiberglass whitewater kayak.

It was a mistake to try surfing my own, cumbersome, PBK-11. Its cockpit was too long and open, and Rottingdean beach ramped steeply with pebbles. The inevitable nosedive swamped the kayak and filled it with shingle. Hauling the waterlogged kayak ashore, I scooped out pebbles by the handful. To my disappointment, dozens remained stubbornly lodged between the tough skin and the wooden frame. There was no way to free them, and it disturbed me to see the rash of bumps across the hull. Thumping the kayak to rattle out the stones proved futile. Most remained embedded, threatening to work their way through the skin like hatching parasites. It was time to look for something more suitable.

So, in 1968, I replaced the PBK with a used, red, fiberglass, whitewater kayak. Craftsmen at Streamlite Mouldings, a kayak

Varnish and Canvas

company in the nearby town of Shoreham, hand-laminated this *KW3*. Designed by Keith White, hence KW, it was a flattish slalom kayak, with an almost rectangular cockpit, and a deck that sloped away to each side from a central ridge.

Competitors paddling KW kayaks won international slalom competitions. These kayaks conformed to the regulation minimum dimensions for competition: thirteen feet two inches long, twenty-four inches wide. At that size, they were not only great for white water rivers, but they were also fun to surf. Members at Brighton Canoe Club (for canoe read *kayak*) mostly used the KW3, KW4, and the new KW7, on the sea.

This kayak club was under one of the Kings Road arches, the brick arches that support the Regency seafront promenade which runs like a shelf along the shore. Completed in the 1880s, with the undercliff arches numbered from west to east, the arches 216 to 224 became home to the Brighton fish market. A hard, between the arches and the beach, served as a stand for wooden beach-launched fishing boats, and extra fish stalls. I remember the smells, and the intense bidding activity at dawn, when my parents took us to the fish market as youngsters, the gleaming hard puddled with fishy water. The market opened daily until 1960, when it moved away for hygiene reasons.

The kayak clubhouse occupied a smaller archway, one hundred yards to the east, squeezed into two levels beside a flight of steps. Kayaks slid end-first from beach level into any available space in a hodgepodge of wooden racks. Once the door shut out daylight and the weather, only a small tough window, through which we could see ankles on the steps outside, shed light into the humid space, aided by a naked light bulb dangling from a twisted lead.

A steep open wooden staircase led up to a social space where senior club members also repaired their kayaks. Spilled polyester resin, drips of colored gelcoat, hard in places and sticky in others, and clumps of matted glass fibers encrusted the bare floorboards. There was always the smell of wet fiberglass, rancid clothing, and curing resin. Adding to the olfactory mix was the occasional reek of fat and vinegar from discarded fish-and-chip wrappers. It was damp too, from what went on downstairs. Since paddlers rarely fully drained their kayaks, or tipped out all the pebbles, before sliding them onto the racks, the ground floor was always pebbly and wet. There, fallen paddles, salty lifejackets, and dripping wetsuits hung from nails created a discouraging obstacle course.

Most members loosely respected each other's gear, although I particularly remember one transgression. Someone was building a slalom kayak from a three-part mold. He had secured the cockpit and seat and had just finished joining the deck to the hull. It looked beautiful, gleaming, pristine, and smelled of fresh resin. Leaving it upstairs for the resin to cure, he warned us that nobody should use it until he had installed the flotation and a footrest.

But the weather turned windy, and the sea got up. All the older, more experienced paddlers turned up to ride the waves. With the kayaks of choice already in the surf, one club member decided to use the new kayak. He said he would be careful to avoid damage so nobody would know he had used it.

Things went astray at the end of a wild ride when the wave reared up into a shore-break, plunging him violently. With his bow planted into the steep bank of flint cobbles, he somersaulted. Without a footrest to stand on, his legs slid down inside the kayak on impact, letterboxing him up to his armpits in the cockpit. The bow, fragile without flotation blocks, snapped off. Bystanders dragged him from the shore-break. He was understandably traumatized, still trapped in the kayak, with his feet hanging out from the broken end.

In 1969 I sold my *KW3* and bought a brand-new, translucent blue, slalom kayak, a *Funa* built by Gaybo Ltd. This one was technically a *blem,*' Graham explained, blemished by thin silver glass fibers visible, here and there, that had not been wetted out completely during lamination. There was nothing wrong with the kayak; it was the Goldsmith brothers kindly finding an excuse to make the price affordable to a young aspiring kayaker.

A Yugoslavian slalom champion designed this *Pavel Bone Funa,* one of the first two designs Gaybo brought back from Germany to build under license. My friend Krista bought the slightly lower volume model, the *Minislalom*.

Krista's parents rented one of the small seafront arches near the West Pier, in which they stored an open boat and fishing tackle. This hideaway became, for a time, our kayak storage space, and a meeting place for me, and Jez, along with Krista, and her younger brother Hans-Peter. After surfing, it offered a warm and cozy retreat. Our kayaks stowed aside; Hans-Peter would fill a kettle to boil water on a camp stove for coffee. Sadly, his hot drinks never tasted as good to me after a mouse, albeit boiled, tumbled from the spout into my mug one day with the hot water.

Hans-Peter inspired me when, one summer, he quietly took off in his sea kayak, returning months later having made a solo circuit of Britain. He brought back tales of selling the fish he caught daily up the east coast. He also unwittingly entered an exclusion zone while a military exercise was underway, successfully dodging the search craft trying to intercept him for hours. His adventure inspired me to make my own, shorter, solo kayak trip in 1975 when I finished college.

Iceland by Kayak

ARRIVAL, FRANCE. FROM LEFT: TIM, KEITH, AUTHOR, ONLOOKER, IAN.
PHOTO BY JAN MCKECKNIE.

3
Sea Kayaks

My early Brighton surfing experiences, even on gentle days, were cold and wet. At low tide, I paddled out through the small surf and turned around to catch a ride. Thrilled each time I surged forward on a wave, I resigned myself to the inevitable broach, when my kayak twisted side-on to the wave and tripped me, throwing me sideways toward the beach into the water.

On emptying the kayak, it was time to paddle back out to repeat the cycle. It did not occur to me there might be a better way. If technique could help, that was slow coming, and not before my first winter surfing in t-shirt and shorts. One priceless lesson was how to tailor a tunic of insulating neoprene to ease the chill of my frequent immersions. Gradually my skills improved.

Discerning surfers do not consider Brighton's waves *surf*. However, riding ragged wind-blown waves, and negotiating the steep dumping shore-break, proved a great preparation for tackling the Atlantic swells of north Devon and Cornwall. The opportunity arose to join friends on an organized summer kayak-surfing trip in 1971. There, intimidated by the height of the glassy waves and anticipating a beating, I was relieved to find I could easily burst out through these swells. Compared to the squat, brute power of Brighton's storm-blown chaos with its jarring impact, these waves were regular and predictable. Each evening, the sun danced through the bottle-glass gleaming walls.

In time I began to see patterns in how the waves' slow rhythm built into sets and subsided into lulls. In one area the waves always looked smaller, easier to paddle out through. Over there, a submerged sandbank changed the direction of the swells, refocusing their energy to form the biggest breaks. Alongside, a rip ran out to sea like a hidden river.

While there, we watched the national surf-kayaking championships, trying to better understand the basics of surfing. Taking to the waves, I still avoided swimming at all costs and strived to guide my kayak in balance at every turn. Unable yet to roll, my paddling followed a tightrope between the relaxed flamboyance I tried to show, and fear of mistake. Watching each competitor's signature style of riding waves was an inspiration, encouraging me to practice, to compete the following year.

At the end of my grammar school education, (High School), it seemed pointless to go to university with no career ideas. How would I choose the most appropriate subjects to study? I needed time to decide, and luck was on my side. My local outdoor activity center took me on as a pre-college trainee instructor, for a year. Earning a meagre £4 per week plus room and board, there was no better way to receive outdoor instructor training with a rewarding social life.

Each week a new group of school children arrived by coach, with teachers. Accompanying them came two physical-education students of about my own age, who came for experience in Outdoor Education. Comfortable with the routine, I lodged in a room above the dormitories in the rambling country house. Between courses, my small motorcycle, a 250-cc single cylinder BSA, or *Beeza*, carried me roaring along the country roads, breathing in the fragrant summer air on my way to visit my parents, and my girlfriends.

Sea Kayaks

That year convinced me to pursue an Outdoor Education career, best approached while young and single. Since a decent job in Outdoor Education required a full teaching qualification, plus a year or two of school-teaching experience, I sold my motorcycle and moved to Bristol to study.

The intimately small college, snugly linked to an adjacent art college, was perfect for me. During the week I helped run a folk club there. A short walk took me to the Avon gorge to rock climb on limestone cliffs. Weekends offered opportunities to hike and climb with other members of the climbing club. Sometimes we hitchhiked to Wales and headed into the mountains, sleeping out under the stars, or in the rain. Other times, with my friend Tim who owned a car, we drove to Devon or Cornwall to ride waves together, entering surf competitions in both slalom kayaks and surf kayaks. Whitewater kayaking drew us too. Joining our friends Keith Robinson, Ian Matheson, and Phil Quill, to compete in whitewater races, we gained access to otherwise restricted rivers.

A piece of advice I took to heart while working toward national kayaking qualifications was to participate in every discipline of kayaking. Each focused on different skills. Marathon racing required balance in a tippy kayak, a good forward stroke, and stamina. Surf-kayaking demanded an understanding of waves. The surf kayaker must lean into turns on a wave and edge the kayak to make the rail grip. Leveling released the rail. Leaning forward or back trimmed the kayak by weighting the nose or tail.

On the other hand, reading river currents and eddies was crucial to whitewater paddling. Whitewater racing combined forward paddling skills with the ability to read the water currents far ahead, to select the fastest line between the rocks down rapids. Slalom was all about accurate blade placement, efficient turning control, and clever use of eddy lines.

I bought my first sea kayak in 1974, from a kayaker in Devon who won it as a prize in a surf competition. "Just look at it!" he scoffed. "I design and build my own surf kayaks to ride waves, and for first prize in a surf competition they offer me that? What were they thinking? For surf?" His agile, flat-bottomed surf kayaks were eight feet long. This *North Sea Kayak,* Derek Hutchinson's first design, was sixteen feet six inches of slender keel. But the antithesis of what he wanted; a straight-tracking sea kayak was just what I was looking for.

Later that year, Jan, instructor at the outdoor center, approached me. "Nige, Ian is planning a trip to paddle across the English Channel. Do you want to come?" Ian Matheson was an Outdoor Education teacher in the Lake District. When I showed enthusiasm, Jan added, "I don't know if I can sit in a kayak for that long. Can we go for a practice paddle together first?" With a reason and opportunity for me to evaluate my new sea kayak too, we embarked on a crazy and memorable weekend coastal trip, in the south of England, where we intended to paddle for seven hours without getting out of our kayaks.

Feeling ready for the Channel crossing, we met with the rest of the group at Dover. Spreading the chart sections, Keith had photocopied for me, over the hood of his hand-painted orange Morris Minor, formerly a Post Office delivery van, I carefully applied transparent sticky-backed plastic to waterproof each sheet. A chart of open water was still as much a mystery to me as a computer punch card, whether viewed from either side. My kayak had neither deck lines nor bungees to hold a chart, so I taped the first section to my deck. Showing the Dover coast, it became redundant the moment we turned our backs to the land.

Afloat, Ian led us confidently from Dover to the red and white Varne lightship, anchored somewhere around mid-crossing, where we drifted. Ian, Keith, Jan, Tim, and I scanned the horizon with nothing else but water in sight. Then, we pushed laboriously to reach the low, bleak sandy beach of Sangatte, France, where the mayor rewarded us with dinner.

Sea Kayaks

Ian and Keith listened to the radio shipping forecasts at midnight, and at six in the morning, before declaring the weather unfavorable for our planned return. Instead, we paddled to Calais, where we carried our kayaks aboard the cross-channel ferry. Standing at the rails and staring back at the wake, Jan suddenly turned to me with a look of dismay. "Nige," she asked, "did you bring your passport with you?"

"No, it never occurred to me. Did you? No? Oops!" We had just a couple of hours to plan how we might circumvent Passport Control, and Customs, on our arrival.

Recognizing how ignorant I was about charts when crossing the English Channel, Ian planned to set that right. That summer he mentored Tim and me on reading charts. He encouraged, and offered guidance, as we planned our first, serious, multiday trip. It was to be at spring tides along a strongly tidal stretch of Dorset coast, around Portland Bill to Shambles shoal and onward. Everything began to come together: the tidal stream atlas and tide tables, walking the parallel rules and dividers across the charts, using the compass rose. Sea charts, with their special abbreviations and symbols, began to make sense. Reassuringly, we experienced on the water what we had predicted in our planning.

When Gaybo introduced the new *Eski* to their range, a sea kayak by the German designer Klaus Lettmann, I ordered one. In the college locker room, I installed fiberglass bulkheads, plastic dinghy locker hatches with screw-lids, and a compass. On graduating from Bristol, I joined Tim in north Cornwall for a weekend of surfing. From there, leaving my surf kayak with him, I set off in the *Eski* to paddle home to Brighton.

That solo trip was like a key to a new door. Joyfully independent, I chose my own route, deciding where and when to land. I slept unnoticed beside my kayak, on the beach, under the night sky, cooking over fist-sized fires of driftwood twigs. But not everything went as planned. I narrowly avoided disaster in a massive set of breakers, close to the Vyneck Rocks, near Land's End. After a hair-raising surf landing, at Gwenver beach near Sennen, I stood on the heavily scoured beach suitably humbled.

Tide races, open crossings, and the wind, worked me until my body ached. Yet on the beach at night, I smiled with satisfaction at the miles covered and the hurdles negotiated. Stretched out on my back on the pebbles, I gazed at the full moon sailing across a starry sky. When my father collected me from the beach at Brighton, ten days later, I felt fit, fulfilled, and hungry for more.

Teaching offered generous vacation time, so next summer I headed north to Scotland with three friends. Dave, an army physical training instructor, was a strong paddler who had come close to winning the grueling 125-miles, non-stop, Devizes-to-Westminster kayak race. We had crossed the English Channel together one night, paddling back the next morning with baguettes, cheese, and wine for an evening party.

He and I, with Keith Robinson, and Helen, set off with a vague plan to circle the Isle of Skye. Since Keith had limited time, we chose to explore the most dramatic sections of coast first, weaving between waterfalls and sea stacks, probing into sea caves, and finding secluded flower-studded camp spots.

Our trip stalled after Keith left. We had picked up two Italian girls who liked to sunbathe topless, so we camped in one spot with them instead of moving on. Each evening, when the surface of the sea roiled with shoals of mackerel, we paddled out to fling bare hooks into the water, hauling up sleek gleaming fish. Back ashore within minutes, we rolled the oily fish in oats, pepper, and wild thyme gathered from the shore, and fried them. It was fun to hang out there, but this was not what we had set out to do.

With time running out, and reined in by our consciences, Dave and I decided to cross the eighteen miles of Little Minch to Lochmaddy, on North Uist, in the Outer Hebrides. When by chance we met kayakers at the mouth of Lochmaddy, Ian Matheson was unrecognizable in sunglasses, hair bleached by sun and salt, not paddling his usual *Seafarer* kayak. "We're circling the Outer Hebrides, with Colin Mortlock. He's gone ahead into Lock Maddy," he explained.

I knew of Colin but had not met him. He had completed a landmark trip the previous summer, 1975, from Norway's Arctic Circle to its northern tip, Nordkapp. With his mountaineering background, Colin staged the event in the style of a climbing expedition. The Sunday Times kept track, reporting with photographs in its new color supplement. The paper hailed it as the first, major, modern sea kayaking expedition, and it was certainly the best publicized. Colin had secured a real publicity coup, and the Nordkapp Expedition inspired many.

Director of the Center of Adventure Education, at Charlotte Mason College in the Lake District, Colin lived for adventure. In his booklet *Adventure Education,* he wrote:

Adventure is a state of mind that begins with feelings of uncertainty about the outcome of a journey and always ends with feelings of enjoyment, satisfaction, or elation about the successful completion of that journey.

Had not my south coast of England trip been such an adventure, by his definition? Intrigued, both by Colin's expedition and his philosophy, I looked forward to meeting him.

We joined him in Lochmaddy, where we took a table at the hotel and drank pots of tea while he reminisced. Inspired, I was in no hurry. It grew late before Colin announced, "Well, we'd better get going. We have our camp to set up before dark. Where are you two planning to camp?"

"Oh," said Dave, remembering that we still had eighteen miles to go. "Our tents are on Skye, so I guess we'd better start back."

The sun set lazily behind us in a blaze of color as we aimed east across Little Minch. Meandering to see the sunset more easily

Iceland by Kayak

over my shoulder, I from time to time called out, "Dave, there's a shark behind you!" He always laughed. Periodically, he shouted, "Killer whale! Killer whale!" We had been at this all week since people told us they had sighted orcas recently, but we had seen neither sharks nor whales.

The light was gradually fading before suddenly, a large black shape surfaced just ahead, and I gasped. "Dave! There's a whale!" He was gazing behind at the fading sunset and turned to me with his usual wide smile.

"Yeah, and look, there's a shark!"

"No, seriously!" I protested, "it was straight ahead! I'm stopping."

"Oh, so you're tired? You need a rest already?" he taunted; his smile wider than ever until something caught his eye. As the broad dark shape arose, Dave's eyes widened, and his smile puckered. I dissolved into laughter.

Another whale blew, and another. Whales surrounded us, eight miles from shore. As more surfaced, I felt both exhilarated, and anxious. What if one ran into us? Dave and I sat together until all seemed quiet. Subdued, we aimed once more for Skye.

Even after nightfall, it was enjoyable to paddle over the sloppy seas toward the darker shape of land ahead. The waves, just enough to push our kayaks around, diverted my attention from any aches and pains I might have noticed on calm water. But suddenly, I became alert and cautious.

"Can you hear that, Dave?" Something was muscling gently at the surface. Torn between passing or making a slight detour to see what it was, my curiosity won. I crept forward, peering into darkness, anticipating a shock if whatever it was took fright, or dived with a splash. My knees gripped tightly, and my hands clamped my paddle. Warily sneaking closer, and closer, even when I could judge the size of the wallowing back, as I took it to be, I could not make it out clearly. I stopped, too timid to continue.

Sea Kayaks

Then, in a flash, recognizing it for what it was: a large wooden cable spool, the flat of its disk awash in the dark, I laughed aloud. Yet, despite my relief I remained on edge.

We resumed, eventually realizing that the lights we saw, periodically flashing in short bursts from shore, were Dave's car headlights. Helen had hit on a brilliant way to signal her position and lead us to land.

On our drive south from Skye, the back of the exhaust pipe parted from the silencer. For the rest of the journey, the Rover 2000 roared like a tank, curtailing any conversation. My mind wandered, recalling my solo paddle from Cornwall along England's south coast. What would it be like to experience sea conditions like that along a less habited coast, with wildlife as plentiful as in the Scottish archipelago? Did Colin Mortlock find that combination in Norway?

With our ears ringing, we at last shut off the engine at Dave's house in Maidstone. There, I pulled a hefty world atlas from his bookshelf and thumbed through to find Norway, curious to see exactly where Colin had paddled. Their trip began at the Arctic Circle near Bodø. From there they negotiated the islands and cliffs of the fjord-cut coast as far as Nordkapp. According to their accounts, it had been a wild, cold, and windy summer. At that same time, I had been paddling England's south coast.

I called out to Dave, who was in the kitchen making coffee, describing what I saw. "Do you fancy paddling in northern Norway?" But, spotting an inset map of Iceland at the top left corner of the page, I had a fresh idea.

"What about Iceland, Dave? Could we do that in one summer?" I looked more closely, adding: "It looks possible."

Dave stuck his head around the door with a quizzical grin and vanished again. I began elucidating, "The south looks smooth, I expect it has beaches along a low coast. The north looks like fjords, so more likely cliffs. Anyway, overall variety. So, what do you think? Shall we paddle around Iceland?" I heard Dave chuckle. "No, really!" I insisted. "I'm serious." But when he set down the coffee, he grinned at me as if humoring my naïvety and shook his head.

THE AUTHOR, ENDING HIS SOUTH COAST OF ENGLAND TRIP.
PHOTO: PETER FOSTER.

4
New Kayak for New Plans

By autumn 1976, working with Keith Robinson on plans for a new kayak, I listed my critiques of all the sea kayaks I had paddled. I knew what I wanted to achieve, and thought I knew how to do it. For one, the cockpit should be farther back, reducing the kayak's tendency to weathercock. The keel, parallel to the water surface, should neither rise nor droop at the ends.

I prioritized a tippy, narrow, rounded hull shape, both for speed and to minimize the wear that always occurred along the ridge of a V-keel. Held upright, a rounded hull was wobbly, so I would add a hard chine for secondary stability. With the kayak held sufficiently far on edge to release the stern for turning, that chine must engage to provide stability.

With slab sides, angling up and out from the chine to the gunnel, the waterline beam would increase with load, adding initial stability. There should be sufficient volume to carry an expedition cargo. Bending my knees comfortably as for paddling, I could measure how high the deck should be at my knee and at my feet. For the ideal spray deck seal, the cockpit should be as small as possible. My list of design criteria kept growing.

Iceland by Kayak

Keith, a wizard with numbers, worked as a computer programmer at University College London (UCL). He streamlined their most popular programs to reduce run-time on their main frame computer. When I visited him in the computer room; a hive of activity, he carried huge flat reels of magnetic tape and boxes of punch-cards. Occasionally he ran the printer, which swallowed yards of thin continuous paper, spitting it out zigzag into a stack. That computer room was like a factory, through which, with Keith, I passed unnoticed.

Keith lived in a tiny basement apartment, just off King's Road in London's swinging West End. He had the convenience of the tube, and bus routes, and places to buy food, all within yards of his door. My own lodging in Buckinghamshire, fifteen miles away, was close to a major motorway, a short walk from Denham, the cozy English village of film studio fame, and near to the last stop of the Piccadilly and Metropolitan lines of the London Tube. I seldom spent weekends in Denham, preferring to hitchhike west, or catch the tube into London for a train to the south coast.

When Keith had something new to show, his distinctive colored Morris Minor van would appear outside my lodgings. Under his arm he carried the now-familiar folded paper sheets, with the spooling holes down each side. I cooked, and we sat on the floor to eat beside the spread papers while Keith pointed out irregularities in the lines of numbers. He deftly circled potential problems here and there in blue biro, inconsistencies of patterns that I struggled to understand.

His later printouts, generated from the numbers, showed profile and plan outlines which I could understand. Over the profiles, I sketched the lines I thought would look better. Increasingly refined, our project progressed on paper until Mac, from the Adur Center in Shoreham, offered to take on the job of building a plug, a solid full-size model of the kayak. It would be a Youth Center workshop project. To start, we cut a series of cross-

New Kayak for New Plans

sectional panels according to Keith's computer printout and mounted them onto a frame of lengthwise stringers. This formed an eighteen feet long skeleton of the kayak.

Keen to work on the plug, the moment school ended each Friday I boarded a train to London, traveling onward via Brighton to Shoreham. On Sunday nights I returned home to mark schoolbooks. Our kayak took shape, in no small part due to the arduous work of enthusiasts at the Adur Center.

Meanwhile, my plan to paddle around Iceland was evolving. The concept was simple: I would start and finish in the same place. But where should I start? I needed more detailed maps.

I mailed a check to the Admiralty Chart agent in London to order charts and the Arctic Pilot, *Sailing Directions*. Excited to receive that weighty tome, and to unroll the charts, I began cross-referencing between them, poring over every detail. Since sea charts reveal little of what to expect on shore, I sought land maps too. Dick Phillips, who ran summer hiking tours in Iceland, sold me what I needed, and offered encouragement.

Now my plans had advanced enough for me to invite others to join me. Keith declined. "My legs get too cold," he confessed regretfully. "Why can't you pick somewhere warm?"

At a sea kayaking symposium, run by the Advanced Sea Kayak Club under John Ramwell, I ran into Geoff Hunter. I knew him through our mutual friend, Jan, and enjoyed his company. In 1974, he made a solo circuit of England by kayak. In an epic along the way, according to Jan, after surviving a night clinging to a buoy in the Solway Firth, he swam four miles to shore. An energetic, cheerful, and resourceful pragmatist, he would be great company, and dependable if things turned bad.

Excitedly, I outlined my proposal. Would he like to come?

"I don't think so," he replied, smiling wryly, "I'm too old for another trip like that." I was disappointed. It seemed a poor excuse: he was only in his mid-thirties. Were there other reasons, or had his solo trip been enough?

"Yeah, well, I am doing construction work at my shop, converting it. I'm building an open spiral staircase out of wood," he explained. "It's pretty tricky."

In contrast, Ian Matheson, keen to come, was confident he could secure the necessary time off work. I visualized a group size of three, adequate for safety yet small enough for joint decision-making, so I needed one more paddler.

According to experienced expedition sea kayakers, I should allow two years or longer to plan such a trip, and I began to see why. Hoping to leave in just six months' time, I did not even know how to get to Iceland. The travel agent unearthed passenger flights to Reykjavík, but what about the kayaks?

"Perhaps you can find a shipping company to take them, or a fishing boat?" He phoned around and learned of a ferry service, but recently discontinued. "It's been sporadic, not running every year," he said. "Sorry I can't be more helpful. Let me know if you want to book a flight. At least I can do that for you."

Noting the name of the ferry company, *P&O Ferries, Orkney, and Shetland Services*, I wrote to inquire. The discouraging reply confirmed that, on reviewing the results of the 1976 sailing season, they would be unlikely to reschedule next year.

So, what about fishing boats? Trawlers used to sail to Icelandic waters from Hull, Grimsby, Cuxhaven, and Aberdeen, between September and May, albeit without passenger facilities. Three long kayaks would make an awkward load for any fishing boat, even if we had them ready in time. Aside from that, there was another hitch. Until this past June, Britain held out against Iceland's newly proposed territorial two-hundred-mile fishing exclusion zone. Britain finally acceded, ending a confrontational, sometimes violent dispute over cod fishing rights. Could British trawlers still visit Iceland?

New Kayak for New Plans

To my rescue came a promising letter from P&O, followed by a New Year confirmation that a Faroese Company planned to run a 1977 ferry service. Better still, they agreed to categorize our kayaks as personal luggage, carried for no added fee.

The ferry would depart fortnightly, throughout the summer, from Scrabster on the north coast of Scotland. We must disembark at Tórshavn, in the Faeroe Islands, while the ferry made a round trip to mainland Scandinavia. Re-boarding for Iceland four days later, we would continue to the east coast town of Seyðisfjörður. I felt jubilant! We could travel with our kayaks, and we knew where to start and finish our trip.

Days later, at the start of the New Year, more good news came on a card from Geoff. He had changed his mind and would like to come. Was my invitation still open?

Meanwhile, the new kayak project hit a snag. The wooden skeleton first complete, the next step was to flesh it out with rigid foam to make a solid kayak shape. This foam began as two liquids which, when carefully mixed in the correct proportions, caused a rapid chemical reaction. Even a small drop of the mixture would fizz and bubble, expanding rapidly in all directions, magically growing like a scary science fiction life form.

No sooner had we stirred together the components than the reaction began. Immediately poured, the frothing mixture expanded to completely fill the voids between the plywood forms. Once the short-lived activity stopped, everything solidified into a lightweight, rigid, pumice-like material that we could easily cut and sand. With the spaces filled, we trimmed off the excess and used the embedded plywood forms to guide our long sandpaper blocks.

The rough shaping complete, we next covered the plug with a fiberglass skin, over which we spray-coated grey car-body-filler. This built up a homogeneous layer that we could shape to symmetry and bring to a polish. Hand-sanding each fresh coat of filler revealed any low spots, which we painstakingly diminished, and eliminated, by respraying and sanding. In time, we created an even enough surface to use successively finer grades of sandpaper. A strategically placed lamp helped highlight and shadow any remaining imperfections.

KEITH WORKING ON THE KAYAK PLUG AT THE ADUR CENTER, SHOREHAM.

New Kayak for New Plans

In winter, as the workshop grew colder, it became evident from the pattern of low spots that the foam was shrinking. Imagine the skin of an emaciated cow, with its ribs jutting out. No matter how often we filled the hollows from the outside, the ribs steadily revealed themselves again. There was no way to arrest the shrinking.

"There's only one thing to do," declared Mac pragmatically one day. "We should make a mold from this before it shrinks even more. We can build a test kayak. If it performs how you want, we'll start again from scratch."

It was December 28th before the test kayak was ready to paddle. The long, slender, red vessel had a tiny cockpit. With no deck fittings or hatches, I thought it looked ugly, but we could change that later. On the water it was fast. In my notes I wrote, "...goes like a rocket and turns quite easily empty." An expedition kayak should perform better loaded than empty, so two days later I made my first test with cargo.

I weighed eighty pounds of shingle into bags. Since the kayak prototype had neither hatches, nor bulkheads, I pushed the bags from the cockpit, balancing the load as evenly as possible between bow and stern. With a winter gale blowing, and aware how quickly the kayak would sink if swamped, I could not risk taking the insecure load out to sea. I stayed inside Shoreham Harbor.

The kayak was fast, easy to keep on track in the wind, but slow to turn with the extra weight. A month later I made a solo trip from Shoreham, paddling east for thirty miles or so in moderate conditions. I passed the white chalk cliffs of the Seven Sisters and Beachy Head, to arrive at Eastbourne in the dark. Returning next day I hugged the shore, where eddies helped ease my work against the tide. Happy with how the kayak performed, I was eager to start building a new plug.

Back to trip planning, I was certain we could buy food in Iceland along the way instead of shipping it ahead or taking it with us. To buy food we would visit villages and meet people, making the trip more interesting. Plus, we would not have to decide in advance what we might want to eat after one, two, five, or nine weeks of paddling.

Although fresh food was best, I planned to take a week's supply of freeze-dried meals with us from England. These, not available in Iceland, would be for the 140 miles of south coast sands, where we would be unable to buy anything. That way we could carry compact food as a backup, to cover for weather and sea conditions which could easily pin us down for a week.

For accommodation, we would carry the nylon rain sheet of a tent, without the inner tent. This would let us set up shelter quickly and easily on beaches. We could change inside if necessary, and cook, leveling the beach material for our mats and sleeping bags later.

Of all the potential kayaking challenges, the south coast appeared the trickiest. The map showed a smooth, gently curving coastline suggestive of sand or shingle. The Admiralty Arctic Pilot Volume 2 described it in more detail:

On the SE coast, between Dyrhólaey and Hornafjörður, 120 miles ENE, there is a narrow but perfectly flat strip of coast, formed of fluvial detritus brought down by the innumerable streams which break out from the clefts and glens of the plateau. These streams bring down great quantities of gravel from the plains, and wherever the surface of the earth is inundated by the ice-cold water, vegetation *refuses to grow. This coast, the SE, is destitute of harbors, all the fjords having been filled up by the detritus carried down by the glacial torrents. Heavy surf rolls in toward the shore, in many places with such violence as to dam back the glacier torrents, so that a string of lagoons has been formed.*

New Kayak for New Plans

The Pilot also describes this coast as being in flux, with rivers overflowing and changing their courses, and fine sand blown by the wind into constantly changing sand dunes. It warns of areas of swamps and quicksand, and powerful currents around the mouths of the bigger rivers. It points out the difficulty and danger of trying to make progress on foot, especially because of the rivers.

The advice for mariners, in case of shipwreck or stranding, especially during the dark of winter, was a warning to us too. Constantly rolling surf on steep beaches with no sheltered places to launch or land? Camping on drifting sand in high winds, with no nearby habitations? At the least we must carry extra food and devise a way to secure our tent.

The next, but a lesser challenge, appeared to be Reykjanes, a peninsula that juts out into the current in the southwest of Iceland. Exposed, it at least offered sheltered places to land.

The north coast promised a different kind of challenge. The chart showing the whole of Iceland marked heavy overfalls and tide races around the major headlands all along the north coast. In our favor, the heaviest action of these tidal rapids should be farther from shore than we would go. The close-to-shore races appeared only at the headlands along the northernmost tip of the northwest peninsula. It puzzled me how the most detailed charts showed no races at all, yet the Pilot cautioned:

During stormy weather, or when the tidal stream is opposed to the wind, heavy races occur off Ritur, Horn, and the headlands between them, and these races may extend several miles seaward: they are extremely dangerous to open boats and even vessels of moderate size are liable to considerable damage when passing through them, and they should therefore be avoided.

Fourth and last of the major challenges might be Langanes, the northeastern point of Iceland, of which the Pilot cautioned:

Iceland by Kayak

Langanesröst, a strong tide race, is often found off Langanes, and may extend far out to sea even in calm weather. It is caused by the meeting of large masses of water brought by the resultant east-going stream along the north coast of Iceland and the SE-going East Iceland Current, the former being increased during the E-going tidal stream.

We would learn more about the other headlands along the north coast when we reached them.

As a schoolteacher, I earned a generous six weeks of summer break, but this trip would take longer. Denied extended leave, I prepared to leave my job at Easter, hoping to find new employment on my return. At the last minute, I saw an advertisement for a teacher/instructor at Burwash Place. It was for a permanent residential post, teaching outdoor education, including field studies as well as activities like kayaking, rock climbing and camping. It was what I had trained for, and I had the required experience, but there was a snag. The job was to start at Easter. I applied anyway and went for an interview. There, one interviewer noted my summer plans and frowned.

I waited while the panel saw the remaining candidates. Then, the secretary called me back to the interview room. "Just to make this clear from the start, we're not offering you the job," one of the panel members informed me as I entered. My heart sank. "Please, take a seat," he gestured. "We just want to clarify one point. I gather you have been planning a kayaking trip to Iceland, for this summer?"

"Yes," I confirmed.

"What I'd like to know, or rather," he glanced to the other interviewers, "what we'd all like to know is, if we offered you the job, I stress again that is not what we are doing, so this is hypothetical, but if we did, would you be prepared to forego your

New Kayak for New Plans

trip to start working here at Easter? You must realize this center needs a full staff through the summer?"

I looked him in the eye. "You know, I would really like to work here," I began, and paused. My mind was already clear, "I would have to decline. I have other people committed to coming to Iceland with me, and they have taken time off, and sponsors have already offered to help us with equipment. I can't let everyone down."

My interviewer stood. "Thank you," he said. "That's all we wanted to ask. All we needed to know. Please wait outside with the others." I closed the door quietly behind me, showing the other candidates, with a shake of my head and a frown, that the panel had not yet decided. I took my seat again.

This was a time of growing teacher unemployment. The Labour government was cutting jobs and trimming budgets. In my final year at college, my professors advised, "You had best take a job if you can get one, rather than spending a fourth year getting higher qualifications. There are likely to be even fewer jobs available by next year. Anybody with a job is hanging on to it."

Having secured a decent job, I had decided to leave it. Now, in with a chance of getting this outdoor education post, I had just thrown away all possibility. I felt tormented.

The door opened, and a face appeared. "Mr. Foster. Would you come in again please?" I glanced at the other candidates and returned to the interview room.

"Mr. Foster, we have decided we like your sense of responsibility. We have chosen to offer you the post on these conditions: we need you here to start at Easter. You will work for two months, and then we will offer you three months' leave; one month paid and two months unpaid. That will allow us to employ someone to cover for you while you are away. After your trip, you will start again in September. We think your expedition will reflect well on the Center and that your extra experience and insight will benefit the students. Will you accept on these terms?" Of course, I accepted and moved to Burwash for the start of April.

The new kayak plug was finally ready, and a two-part mold prepared in time for us to build three kayaks, for Geoff, Ian, and myself. With humble reference to my scary solo experience near Land's End, we named the new kayak the *Vyneck*, after the strangely named rocks there.

We readied the kayaks, creating deck anchor points by threading small loops of rope through holes, drilled into the deck, splaying the ends and fiber-glassing them flat under the deck. The loops above secured our deck lines and bungees. We carefully positioned the front bulkheads to double as foot braces, maximizing the cargo compartment space and minimizing the cockpit volume.

With few reliable options for hatches, we laminated hatch rims using the top of a bucket as a form. The rims sat recessed into the front and rear deck, bringing the lids flush with the deck. For hatch lids we used nylon fabric, stuck onto plywood disks to prevent implosion. Each lid sealed around the rim with a rubber band cut from a tractor tire inner tube, plus a strong bungee.

I mounted my compass on the front deck and installed a hand bilge pump behind the seat. The pump's bulky chamber hung below the deck, just behind the cockpit. I could work the handle, above deck, without unsealing my spray deck. The intake pipe lay beneath the seat.

New to fitting a pump, I worried that if I drilled through the side of the kayak, water would run in. Instead, I positioned the outlet on the deck, within reach behind me, where I could seal it with a stopper. I had no idea how much I would regret my decision.

Another detail concerned me. A spray deck seals much more securely around a small cockpit than a large one. Besides keeping the cockpit drier, it reduces the chance of collapse under a

New Kayak for New Plans

breaking wave. So, I designed a small cockpit opening; too small I discovered. Sitting on the rear deck and sliding forward, my legs jammed before I could drop into my seat. My legs were too long. Had the cockpit sloped more; a lower rear deck or higher foredeck, or been longer, then I could have slid in. The others could, with their shorter legs.

The only way I could get in was to first feed my feet in behind me, to kneel on the seat facing back. From that position I could corkscrew in. There was no problem once seated, but to exit again I must first twist to kneel facing the stern, before peeling my legs free. I wondered how that procedure would pan out in Iceland.

It was getting close to our departure date when, for family reasons, Ian made the difficult choice not to come. It was a disappointment no doubt for him, but also for us. At this late stage I chose to continue with just the two of us, rather than trying to find a last-minute substitute.

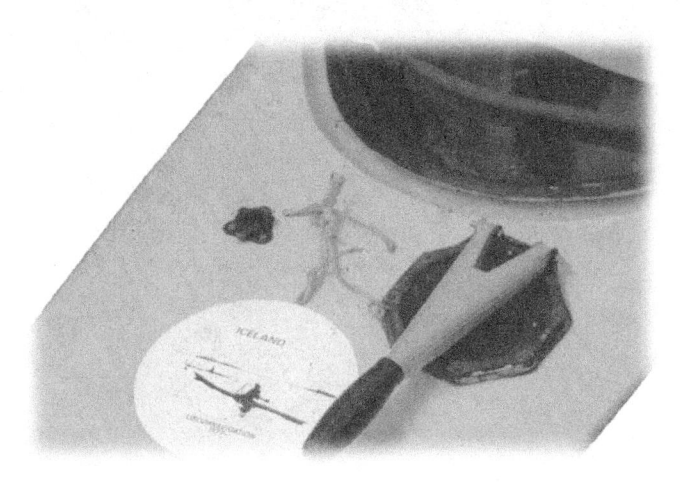

PUMP MOUNTED ON REAR DECK WITH OUTLET, LEFT.

BURWASH PLACE OUTDOOR CENTER, SUSSEX.

5
Tie the Paddles

On the fourth of June 1977, England was preparing to celebrate Queen Elizabeth's Silver Jubilee. Three days ahead of time, in readiness for the officially sanctioned street celebrations, buntings and flags adorned the streets in villages and towns across the country. When I walked to the home of the head cook of Burwash Place, invited to a farewell celebration on the night before my departure for Iceland, decorations hung all along the main street of Burwash village. Everyone was in festive mood.

The next day, Geoff pulled up in the driveway of Burwash Place in his dark green Austin minivan. I loved that vehicle model. It first appeared in 1965, a tiny vehicle with a bare-bones unlined interior. Its diminutive bonnet, *hood*, hid a tightly packed transverse engine. Pulled by that 1,000-cc power pack with a low gear ratio, it was the smallest of panel vans. Having no windows in the back except on the twin, side-hinged, back doors, and no lining, the small vent that folded open in the roof was essential to manage the inevitable condensation.

Iceland by Kayak

Fitted with front seats only, a long-angled stick shift, loops of cable to unlatch the doors, and sideways sliding windows in the doors, it was the most basic vehicle; characterful and fun to drive.

This was Geoff's builder's truck, and since I did not own a car, it was our only option for the journey north to Scrabster. I was sure it would suffice, until he opened the back doors revealing the space already full of his own gear. Where would mine fit? "Well, I do need to organize a bit," Geoff admitted, frowning as he looked inside. He scratched his head, which stood his hair on end, and began unloading onto the lawn.

My gear lay in disarray on the same huge lawn. "I'm not ready yet either. There's no way I could fit all that stuff in my kayak, even without food." Fortunately, any surplus could go back into my tiny attic lodgings. When, despite paring down, there was still too much, I gave up and stuffed what I could into the van. I would fine-tune later.

Lastly, we tied the kayaks onto the roof rack. Each kayak measured eight feet longer than the van, so overhung both in front and behind. Our paddles lay on the roof too. We would carry them as spares, picking up new ones from our sponsor, *Lendal Paddles*, on our way north. Unfamiliar with their asymmetrical spoon blades, I took my usual one-piece wooden slalom paddle just in case. How to carry this 210cm (seven-feet-long), feathered, *Mark G* paddle on my kayak remained a puzzle.

We climbed down into our seats, and I strapped myself in, ready for the long journey. The car exuded the distinctive odor of the engine, warmed by the sun, blending with that from our gear in the back. Geoff glanced sideways at me, then checked his mirrors. Jiggling the long gear stick into first, he eased the clutch, and we were off, turning carefully into the lane. I felt excited. Our adventure had begun.

Tie the Paddles.

MAP 2. THE ROUTE NORTH FROM BURWASH TO SEYÐISFJÖRÐUR.

Iceland by Kayak

As we reached the top of the hill and turned toward Burwash village, Geoff laughingly announced, "Look, they've hung up flags for us! What a send-off!" It was absurd to imagine these decorations, the colored buntings, flags, flowers, and wreaths, were to celebrate the start of our adventure. Yet although nobody stood in the street to wave us off, it was easy to imagine it to be true. We were exuberant, until a clatter in the street behind made Geoff stop abruptly, eyes to the mirror.

"Oh no! We forgot to tie the paddles!" We ran back for them, and I gave them a quick look over.

"They seem all right, just a few scratches," I said, relieved. "We were lucky. Not a good start."

"Just as well we were going slowly."

With such a long drive ahead, we lashed the paddles firmly beside the kayaks, and double-checked the security of all our ties. Satisfied, Geoff turned, his face serious.

"All okay now? Take two?" He broke into a grin. We jumped back into the van and drove on.

Having planned to stay the night at Ian's house near Hexham, in the north of England, our late start meant this longish drive ran into nighttime. We had been on the road for hours when suddenly the fan belt snapped. We slowed to the roadside to consider our options before proceeding, at a crawl, until we located a phone box to call the AA, *Automobile Association.*

We waited. Eventually, a familiar little yellow service van pulled up, with a replacement fan belt. But in the dark, we cannot have tightened the new belt enough, for the engine overheated.

Stopped for the engine to cool, we tightened the fan belt and waited until it was safe to open the filler cap. A nearby river supplied water to refill the radiator. Cramped from the drive, and tired, we debated staying there to sleep. Instead, we pushed on, arriving late to find Ian still wide awake. He was in good spirits and welcomed us in, where we talked into the early hours.

Tie the Paddles.

Next morning, Alistair Wilson greeted us at the *Lendal* workshop, where he presented us with our new paddles. The green and red colored fiberglass shafts looked great, with blades carefully laminated from layers of wood veneer. I had seen Geoff's all-wood, Lendal racing paddle, with its hollow wooden shaft, but these looked more modern, high-tech, and stylish. We tied them to the roof with our others and continued north.

It was not long before we were farther north than I had ever been, speeding along narrow deserted roads, between mountains, across moorland and over tumbling rivers. The landscape seemed bleak, empty, wild, and endless. When we stopped for a break the acrid smell of smoke, from smoldering peat fires in nearby cottages, clenched at my nose and throat.

The drive seemed interminable. The villages appeared fewer and farther apart, until, on a road threading over desolate moorland, we ran out of petrol. As the car trundled quietly to a halt, Geoff frowned and rolled his eyes. We glanced at each other for a long moment before bursting out either side. Standing beside the van, I heard nothing but the sound of the wind, the drifting cries of curlews and golden plovers, the drumming of a snipe overhead, and the clicking of the engine as it cooled. There was not another car in sight.

"Well," I said, "that's it then!" I looked around at the exposed dark bank of peat at the verge. Beyond, the bobbing heads of cotton-grass hovered like a gleaming layer of silver mist above the bog. Where might the nearest phone box be, or the nearest filling station? "I suppose we can try to hitch a ride if a car comes. There's no point in walking is there?"

"Not really. Have we got anything we could fetch petrol in?"

At that, I suddenly remembered the bottle with petrol for my cooking stove. Although only a pint or two, it should get us a little way. I rummaged through the back of the van to find it and emptied the petrol into the tank. The car started, and with whoops of joy we were away again, driving gently to get the best possible mileage. We made it all the way to the next filling station.

On reaching the small port of Scrabster, our destination, we first made a cursory reconnaissance of the harbor to see where the ferry would dock. Then, since today was Geoff's birthday, upon spotting the Ferry Inn on the waterfront we headed for the bar. We scarcely had time for a celebratory pint of Guinness, with a whisky chaser, before the bar closed. Back outside, disoriented by the daylight, I realized closing time was earlier here at 10pm, and we had gained extra evening light by driving north.

Driving onto the moor outside Scrabster, we parked the Mini beside the road and uncovered our sleeping bags. "It's freezing!" I exclaimed. "What a change from yesterday morning. I had my shirt off, and now…"

"I'm looking forward to my bag," broke in Geoff, yawning. "Do you need anything else, or can I lock the van?"

We climbed over the low stone roadside wall, in search of shelter from the wind, where I spread my bivvy bag. I had sewn this red waterproof nylon sack to serve both as a groundsheet for the tent, and as an emergency shelter. Tonight, I unrolled my yellow camping mat, and spread my sleeping bag over it. It was such a simple pleasure to curl up warm, with fresh air breezing across my face. And it was easy to get organized since it was still daylight. "Yes!" I called out to the world at large. "Yes, this is the life!"

When rain woke me, it was still light or already light. I struggled clumsily to get inside the bivvy bag without getting out of my sleeping bag, pulled the waterproof nylon far across my head, and went back to sleep to the patter of raindrops.

"Good morning, Geoff. God save the Queen!" It was Monday morning: Jubilee Day. Although February 6th had been the 25th anniversary of Elizabeth's accession to the throne, June was a better time of year for public celebrations. Official festivities

Tie the Paddles.

would begin today, with the Queen lighting the first of a string of beacons across the kingdom. Sadly, by the time the street parties began, we would have left the country.

God save the Queen was also the title of a recently released song by the Sex Pistols, lyrically criticizing the Labour government's 1970's treatment of the English working class. Released a little more than a week before, on May 27th, sales of 150,000 copies rocketed the single into the charts at number two. It got so much bad press, the BBC banned it four days later, guaranteeing extra sales.

"God save the Queen!" echoed Geoff. "Now, what's for breakfast?"

Assembling the stove on the grass between the car and the dry-stone wall, I fired it up to heat water. Bacon began curling in the pan, spitting, and sizzling. Eggs clouded to white, finally crisping golden around the edges. As the petrol stove roared, the tantalizing aroma of breakfast wafted in the fresh air.

Cooking complete; I shut the stove off. The abruptness shocked my ears. Without the confining roar of the flame, the space around me seemed to suddenly expand. Birdsong warbled in the distance, and the breeze whispered all around.

I dug into the mouth-watering breakfast of buttered bread, with bacon, and egg on top, even tastier beneath an open sky. Yet that, even with coffee, was insufficient to warm me.

"Brr! It's ca-ca-cold," I complained, jiggling my shoulders. "But at least it stopped raining. Did you stay dry enough last night?"

"I was damp, although not too bad," Geoff replied. Then, ever practical, "You know, we'd better sort out what we're going to take. It'll be easier to focus if we do it here, with nobody about."

With the kayaks offloaded we began emptying the van. "Let's pack the kayaks with what we're taking," Geoff suggested. "I can

Iceland by Kayak

drive down with them loaded if I take it easy. It's not far, so long as the roof racks can take it."

From the back of the van, Geoff pulled a wooden trunk which he dropped onto the road beside the rest of his stuff. "I thought we might take this with us to Seyðisfjörður," he turned to me for approval. "We could find somewhere to leave it there, with, you know, stuff we don't want to carry, like a change of clothes and the main repair kit."

"That's clever. I might leave clothes in the van too for the drive home." We packed the trunk and the kayaks. Stuffing the surplus into the van, we slid the trunk in last and closed the doors on it all.

"How can you possibly know what you're going to need?" I asked rhetorically, standing back to see how the van balanced with such a weight of kayaks on top. It looked no different, although it might feel top-heavy to drive.

Our next urgent task was to find safe parking for the car for the summer. The officer at the Scrabster police station suggested the harbormaster, who sized up the car with pursed lips. He escorted us behind his office, to a garden with a shed little wider than the car. "There," he pointed, "you can leave it in there." Turning to Geoff he added, "Of course, we shall make a small charge." We agreed the terms, although we were not ready to leave the car just yet.

The parking area for the ferry, at the base of a crumbling, vegetated, one-hundred-foot-high cliff, was mostly empty. Having parked as close to the front of the ferry line as we could, I stood and compared our tiny car with the other vehicles gathering there. A notable number of Land Rovers waited, with Volkswagen camper vans fully loaded with supplies, and commercial vans.

There was a buzz of excitement. Others too were heading for adventure. Yet, despite my enthusiasm, I felt uneasy. It seemed impossible that I was here. The idea I forced into motion months

Tie the Paddles.

ago had somehow gathered unstoppable momentum, as if the plan had escaped my hands. With a life of its own, it carried me along, powerless.

The tortured wail of bagpipes gathering wind, jolted me from my thoughts. There on the quay stood a compact group of bare-kneed, tartan-clad pipers. Behind them, already entering the harbor, came the ferry. There would be a piped welcome for this first ferry of the season. Geoff ran up the hill to film the scene with his borrowed Bell and Howell, eight-millimeter cine camera. The ferry looked small. It turned swiftly and reversed to the dock.

Waiting for the loading ramp to come down, we were the first to board, driving where the crew directed, all the way to the bow to unload the kayaks. The car deck rapidly filled with diesel fumes from forklift trucks carrying pallets of cargo. We heaved Geoff's trunk from the van, closed the doors, and dodged off the ferry before other vehicles could block us in.

Back at the harbormaster's, Geoff inched the van forward into the narrow shed, shut off the engine, and crawled out through the back doors. We closed the shed behind, trusting the car would still be there on our return.

Re-boarding the ferry, *M/S Smyril*, Faeroese for merlin, the small falcon, we climbed to the upper deck to sit in the sun. Hungry, we keyed open the lids from cans of sardines, broke apart a loaf of bread, and ate. Then, with the ship's horn booming and the bagpipes barely audible above the engine, the ferry eased steadily from the dock toward Pentland Firth. At last, we had begun the first ocean leg of our journey north. Our next stop, the Faeroe Islands, would be in fourteen hours or so.

Smyril,[1] at three hundred feet long, seemed small for the heaving ocean swell that day. One passenger confidently told me that the ferry, built for calm Danish waters, had no stabilizers. "It's not really up to this," he proclaimed. I hoped he was wrong.

Iceland by Kayak

I needed stabilizers. Clutching the handrails against the ship's pitch and roll, I followed Geoff down to the car deck to fetch our notebooks, and to rearrange things. Our plan was to leave all we might need on board easily accessible.

Everything there was in motion. Heavily laden vehicles lifted to the top of their suspension units, and sank back, swaying bodily from side to side. Discordant haunted-house groans, from every vehicle, protested above the insistent thudding of the ship's engines, and the whoomph and shudder each time the ship plunged through the swell. The ship's motion fed the shrieking cacophony like a whip across tortured backs.

Reaching our kayaks, we realized we should have been here before we sailed. It was too late now; the pitching and rolling of the ship foiled us. From the bottom of each trough, the fast-lifting bow forced me down onto the deck. When it reached its high point and began to drop, I became weightless, legs straight, losing contact with the deck.

"Maybe sardines weren't the best thing for lunch," I remarked drily. Geoff looked at me and laughed. Neither of us could control our movements. The only option was to escape from the car deck.

Since few passengers had embarked on this first sailing of the season, Geoff and I had our choice of empty reclining seats with a view of the ocean. Mostly sleeping, I awoke to see the southernmost cliffs of the Faeroe Islands. Despite wanting to see more, I was unable to stay awake for more than a minute at a time. I remember seeing misty islands. White blots spread where the massive swell disintegrated into spray against towering walls of layered rock. Besides that, and myriad seabirds, I registered little else.

The bustle of people walking past alerted me to our impending arrival. Geoff, no longer in his reclining seat, lay nearby, curled in sleep on the floor. We stumbled up onto deck to watch the ship approach Tórshavn.

Tie the Paddles.

SMYRIL APPROACHING FERRY DOCK, SCRABSTER.

SHELTERED PASSAGE, FAEROE ISLANDS.

TURF-ROOFED HOUSES, TÓRSHAVN.

Tie the Paddles.

FORT SKANSIN, TÓRSHAVN.

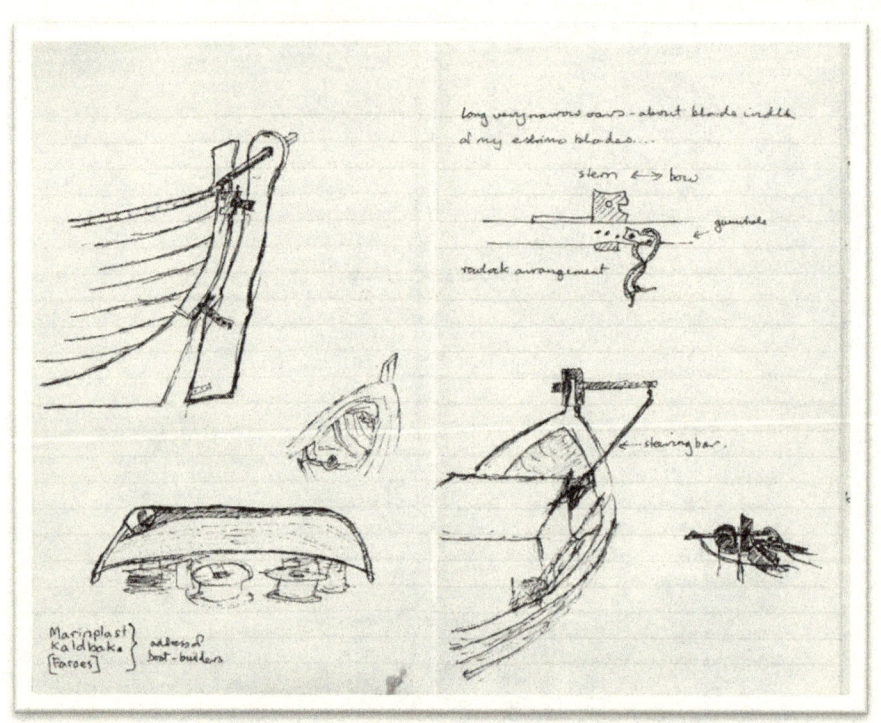

SKETCHES OF FISHING BOATS, TÓRSHAVN.

6
Faeroe Islands

The Faeroe Islands stood steep, and grey, in the wind and low clouds. Closer scrutiny revealed bare, green, treeless slopes and precipitous rock cliffs. Screaming gulls surrounded the ferry decks, gliding in on the air currents to hang just feet aside from the crowded rails where passengers excitedly surveyed the scene. Puffins and guillemots winged energetically from the ferry's path or dived at the last minute. I could see their dark shapes rocket through the water from the hull until the bow wake hid them from sight.

The ferry docked at Tórshavn, a colorful town that spread from the focus of the dock toward the surrounding low hills. Rust-red buildings with roofs of copper green nestled between others with turf roofs.

Geoff and I hurried down to the car deck, which echoed with the bang and slam of truck doors. Growling motors rapidly filled the air with a greasy haze of diesel exhaust. We made haste, trying not to breathe deeply, weaving past drivers and vehicles, dodging Land Rovers; anxious to escape the fug. As trucks took turns to maneuver away, other vehicles raced aboard to fetch container trailers. Between them all, forklifts spun and darted, nimbly managing laden palettes. As hospitable as a hornet nest, this was no place for us pedestrians.

Geoff spotted a trolley and asked a crew member if we could use it to carry our kayaks, saving us arm work. Wheeling each item in turn, we skirted the fish-stinking water, which pooled beside a container on the ferry deck, to escape into the open air. Finally, all lay safe from the traffic.

"Let's ask if it's okay to leave our kayaks somewhere down here," Geoff suggested. When the ferry returned from Norway in four days' time, we would re-board from this same dock. The shorter the carry today the easier for us later, especially considering the weight. Probing into each building along the quay, we found the harbormaster. "If you leave them there, they could get broken," he warned. "Nobody will steal them, but damage? Look," he glanced around, searching for the right person. He pointed, "See that man over there? Christopher. He might be able to help you."

Christopher was busy. "Wait half an hour and I'll find somewhere," he said brusquely. As Geoff and I walked away along the quay, the wind cut so cold I withdrew, walking stiffly, numb to conversation, and loath to wait.

On our return, Christopher called one of his team to bring a forklift. Ferrying our kayaks and box to the P&O warehouse, he set them aside where they would be out of the way.

"Will it be okay for us to come in here to fetch things?" Geoff asked, concerned they might not allow visitors.

"Come any time you like, for the kayaks, or your gear," he said. "We lock up at six every evening and don't open again till morning, but apart from that we're always open."

"Great," said Geoff, "we'll leave our stuff here until we find somewhere to camp." Taking only valuables with us, we set off into town, exploring between tightly clustered, colorful houses with pink, green or red roofs. The older buildings were turf roofed.

Faeroe Islands

I looked at my watch, shocked to see it was six-thirty. If the warehouse locked up at six, we were too late. Then, it dawned on me that it must be six-thirty in the morning, not evening as I had assumed. "Maybe it'll start getting warmer soon," I hoped.

We recognized someone from the ferry, a Belgian, sitting on a bench in a small public garden. He had come to birdwatch in the Faeroe Islands. Two anthropologists from Durham, England, shared an adjacent bench. They planned to spend a couple of years in Iceland. We had joined them to snack when Geoff pulled out a long knife and began slitting the wrists of the brand-new paddling jacket he wore.

His action was so unexpected, his blade so impressive, that everyone stopped talking and stared. Geoff, suddenly conscious he was the center of attention, looked up for a moment and grinned. "The wrists are too tight," he explained, holding his knife point-up in front of him before refocusing on severing the stitches.

Geoff and I left to explore more of the old town. Here were steep narrow passages between houses. Vertical creosoted wooden planks clad the walls from two or three feet above the ground, the dark contrasting the exposed whitewashed stonework below. Simple latticed windows, with white painted frames, lay flush with the walls, under eaves overhung with ragged grass from the turf-covered roofs. Beneath the sod of the roof protruded the gleaming edge of a waterproof underlayer of birch bark, and beneath it under the eaves, dangling like bones at head height, a row of fish hung by their tails to dry.

Aimless, both of us weary, Geoff suddenly suggested, "Let's check out the lighthouse." The tall red and white banded tower stood on top of a small hill, at the east side of the harbor. Capping the hill was a grassed area, with embankments, fortifications, and cannons. According to the information board, this was Fort Skansin. The ramparts offered a great lookout over the harbor, where dozens of small wooden boats moored.

Iceland by Kayak

The fort, built in 1580 to protect against pirate raids, and extended in 1780, has served well. Two guns, from the Second World War when the fort was a British military base, stood on display with older Danish cannons. This was the perfect place for a lookout, a fort, and a lighthouse.

Across the water, a steep green island, Nólsoy, guarded Tórshavn from the worst of any easterly swell. Nólsoy looked bleak, treeless but grassy, with bands of rock exposed on the slopes, although when I glanced around me, I saw the landscape was not so different on this island. But the grass at my feet was short, the ground flat. Might this be a good place to camp? Sheltered too, it was almost ideal.

Geoff agreed. With our tent pitched inside the grassy ramparts, we slept through a large part of the day. Waking to fetch more items before the warehouse closed for the night, we prepared our first hot meal on Faeroe. Then, we began tackling unfinished tasks, such as hand-sewing closures onto our drybags.

With no plans for our days here, we had time to paddle, launching into the small-boat harbor between the wooden fishing boats. Joyful to be afloat, we turned north from the harbor, into the wind and swell, and paddled as far as a tiny harbor, where a flock of eider ducks sheltered, cooing gently. Open clinker-built boats rested on a concrete wall there. Besides a handful of houses and huts, there was little else. Geoff dropped a line to fish, reeling out until he found the water too deep.

Turning, we flew back with the wind and tide. The kayaks sprang to life, racing down the waves and turning easily, although my new paddle almost tripped me more than once. The blades curved from end to end, like my old ones, but also curved from side to side, like a spoon, changing how they caught the water when I steered. I was also still learning how my new kayak performed, realizing I enjoyed it more on the choppy water than in the shelter. But was my seat too wide for full control? Should I pad the sides?

At the warehouse I was crestfallen to discover that mice, or rats, had chewed into our plastic wrapped freeze-dried meals spilling the contents. These meals were to have fed us along Iceland's south coast sands. We would be unable to replace them, so we sifted through the bags to salvage the one or two that were undamaged before ditching the rest.

In the following days we tried to catch fish on hand lines dropped from the quay of the inner, small-boat harbor. I relaxed, sitting on the wall, writing letters, and admiring the fishing boats. The commonest designs resembled little Viking ships, clinker planked, with perky upswept bow and stern, and an over-stern rudder controlled either by a long offset tiller or by lines. The larger ones had a small cuddy amidships, and an inboard engine. Not all carried a short mast at the stern, where a small sail might hold the boat bow-to-wind for drift fishing. The open boats, with inboard or outboard motor, or lacking both, commonly had a simple peg and rope loop as a rowlock, for rowing.

On our final night we could hear music come and go in the wind. It was the Danish pop group, *The Lollipops*, performing at the football stadium a mile away. In anticipation of the next ferry ride, we took a long walk, returning in time to warm ourselves in our sleeping bags, clasping hot drinks, before collapsing the tent. Soon, everything was ready, close to where the ferry would berth.

Aboard, under way, we descended to the car deck to fetch things from our kayaks. It was a dance, as before, trying to balance against the pitch and roll in the bow, as the ferry pounded through the swell. We struggled for balance like drunkards, amid the clamor of engine noise, the crash of the hull against the swell, and the creaks from rocking trucks. It was challenging to stay there long enough to find what we needed, but when we staggered

Iceland by Kayak

between the swaying vehicles toward the stairway, we saw we were not the only ones there, and not the only ones in difficulty.

Three people leaned over a red Mercedes car, which had hitched itself, fender over fender, to the truck in front of it. The owner, a most attractive young woman, introduced herself as Salóme, from Iceland. Together we managed to lift her car free, roll it back, and immobilize it while Salóme scrambled in to apply the parking brake. Mission accomplished, we climbed upstairs, chose a table in the dining area, and sat together.

"Where are you going?" I asked Salóme.

"Reykjavík, I'm going home, how about you?"

"We're planning to kayak around Iceland."

"All of the way around?" she looked astonished. "You'll never make it!" She looked us up and down, and then offered, "If you make it as far as Keflavík, give me a call. I'll treat you to a meal and show you around town." We wrote down her phone number and assured her confidently that we would call her the moment we got there. Of course, we had no idea if we could get that far. What would we make of that long southern beach? Would the surf prove too big? Would we give up in the face of sandstorms, or quicksand?

"What about you, where have you been?" I asked.

"Morocco, trading furs. I've been away for quite a while, so I'm looking forward to coming home at last. She laughed, adding, "But I have a pet mynah bird I'm bringing back. I hope they let me bring it into Iceland, or that customs won't notice I have it." She paused before adding, "I hope my parents like him."

One of the men with her pulled out a pack of playing cards and punched them from the box to shuffle. "Do you want to play?" Salóme invited. I declined and left the four at the table, to find somewhere quiet to sleep. When I returned later to buy breakfast, they were either all still there, or back there again, so I joined them for the rest of the journey.

Faeroe Islands

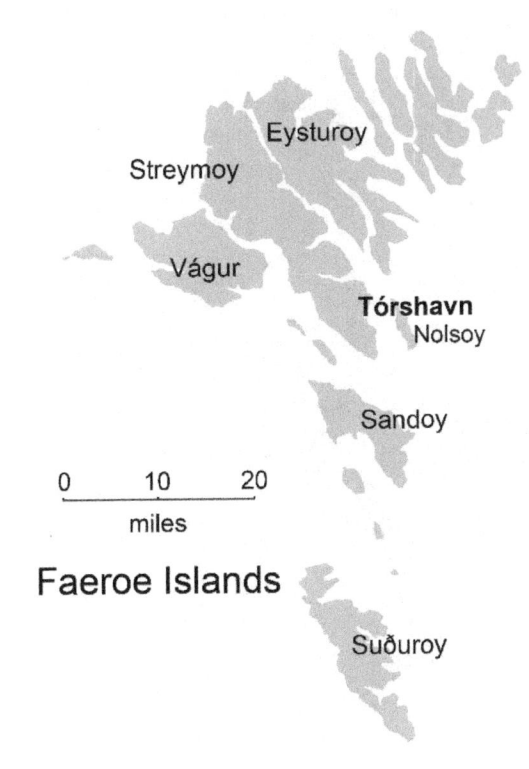

MAP 3. FAEROE ISLANDS.

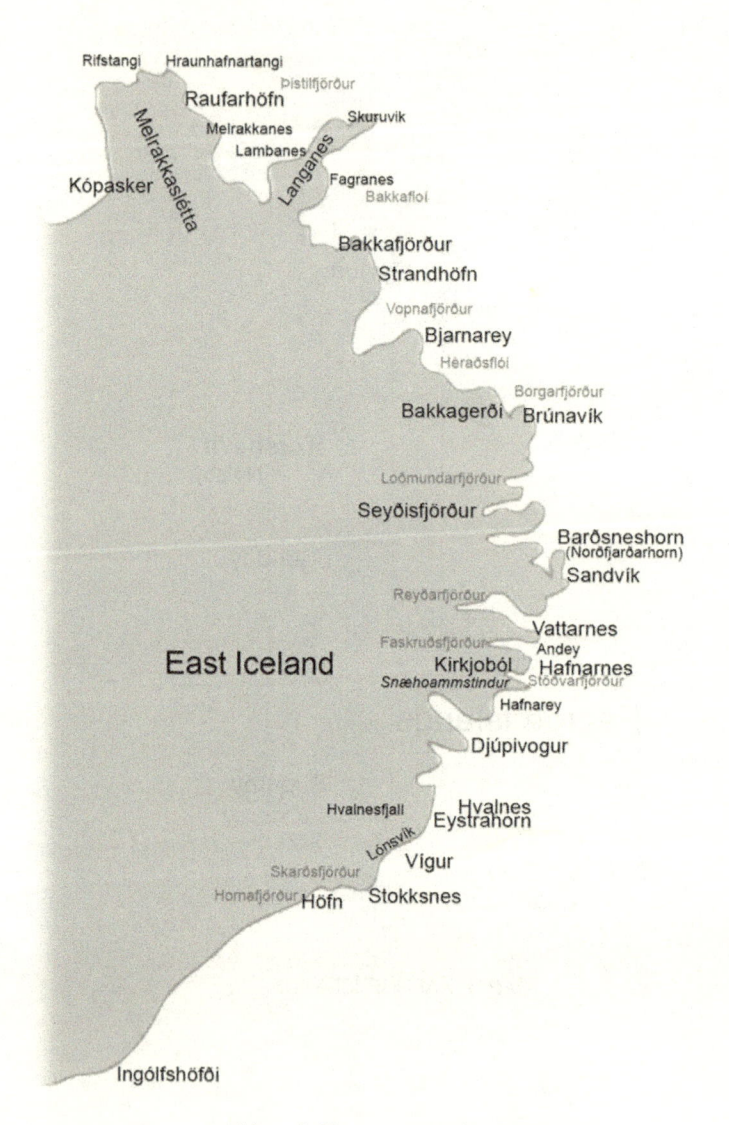

MAP 4. EAST COAST ICELAND.

7
Seyðisfjörður

There was a flurry of activity on board Smyril when the snow-covered mountains of Iceland appeared. We joined the crowd on deck to see. Looking around at the eager faces I wondered, was no one else apprehensive, like me? The anoraks, wool shirts, and leather hiking boots suggested it was mostly the Land Rover crowd up here, swaying with the deck, hands deep in pockets. I felt unsure of my decision to come to Iceland to paddle.

The ferry turned, entering a wide fjord, which seemed to reach endlessly inland between steep, snow-splattered cliffs and crags. The mountains either side, peaking above three thousand feet, looked harsh and rugged, foreboding. Leaving via this fjord would be the start of our paddling adventure.

After the ferry docked that morning, near the end of the fjord and a half-mile from town, customs held us until they finished processing all the vehicles. They issued us a temporary import permit for each kayak, with duty payable only if we did not take them home with us. From customs, it was just a short carry to the closest rocky beach where we might launch.

Iceland by Kayak

On the ferry, and clearing customs, Seyðisfjörður.

Seyðisfjörður

No sooner had we set the kayaks near the water's edge than inquisitive children of all ages ran to see what we were doing. Those recognizing we spoke English, tried their language skills. The little ones simply questioned us in Icelandic and looked puzzled when we could not answer. Relentless, and all around us, they were off-putting. "This is impossible. I can't even think," I admitted to Geoff.

"Yeah, distracting, aren't they? Why don't we go over the road and cook something? They'll probably get bored and go away."

We quickly selected ingredients from our food bags, retrieved the stove and fuel, and retreated to a quiet spot across the unpaved road. Sitting among the rocks, we cooked a quick meal. Having escaped the children, we greeted two girls of about our own age, Marie, and a Swedish girl, who had spotted us and shown interest. They readily accepted our offer of coffee, so we sat together, relaxing in the sun as we told them our plans.

I felt so alive in that fresh air with its sharp chill edge. However, I felt uncomfortable revealing pride in our ambitions. It seemed a little phony since we had not paddled a single stroke yet. It was like a meritless boast with which we could easily jinx our trip. We must make a start. So, with our stomachs satisfied, we strolled into town to find the harbormaster.

"Hjalmar," he introduced himself, extending his hand. We explained our situation and asked if we could leave our spare gear in his care. He was a genial man who seemed interested in our plans, and he said it would be no problem. "You can leave those things in my warehouse."

For me, deciding what to leave was as painful as discovering it would not all fit into my kayak. This was my first experience loading for a major trip. I had only my south coast of England and Scottish trips to draw on. On the first of those, my kayak was half-empty. With neither tent, nor stove, I carried just enough for an

Iceland by Kayak

English summer excursion. In Scotland I carried little more. For Iceland, I had planned for every foreseeable eventuality, except that of not being able to fit everything in.

"Do you think I'll need a down jacket?" I asked Geoff. He turned to consider.

"Well, I think it'll be cold, don't you? Can you fit it in?" I was unsure. The jacket was warm and pillowy, but even when packed tightly into a drybag, it was very bulky. In the end, I took the jacket and left other things I thought I could more easily manage without. Why were all these decisions so tough, I wondered?

As an emergency backup, we left our main repair kit with Hjalmar, assuming we could fetch it somehow if we had a major breakage. I strapped my one-piece slalom paddle, as a spare, to my back deck. It just fit, with one blade flat on the deck behind the cockpit and the other angled against the hull at the stern. I hoped it would not affect my steering. Convinced that more was better than less, that was not the only thing I strapped to the deck.

In Seyðisfjörður, a town of about 1,000 people, news of our departure spread rapidly. When we left to paddle down the fjord that evening, we had an escort of cars honking their horns for all the few miles until the end of the road. We cruised past large wooden scaffolds. Fish hung there to dry, in rows, streaked cream and amber, as twisted and gnarled as mandrakes.

Having seen the fjord from the deck of Smyril, I saw it differently from sea level. The water surface reflected the mountains in lazy patterns of dancing ameboid shapes. I could smell the fish on the drying racks we passed, and the cold tang of seaweed, the air chill in my nose. Although it would take us nine slow miles to reach the end of the fjord, my heavily laden kayak slipped easily enough through the water. Once up to speed, all that first-day weight slid onward, gliding smoothly, even when I paused my strokes.

Seyðisfjörður

When we turned to cross toward the south side of the fjord, the sun dropped behind the mountains casting a cold shadow on us. The sky stayed red, while it got no darker. Around midnight I landed on a small beach, for a pee and to find the water bottle I thoughtlessly packed inside my hatch. I was excited to realize the sky was still light, brighter than twilight. I could see to read the details on my chart with ease, even here in the shade of the mountains. This was a new kind of night paddling for me.

I launched again to find Geoff drifting and fishing. "I've just seen a porpoise," he announced proudly, straightening his back with a smile. He began pulling, winding his line from the water onto a flat piece of wood. When he reached the end, he stabbed the empty fishhooks into the wood and tucked the slab and sinker under his deck bungees.

I was still scanning the sea in vain for porpoises when Geoff called, "Ready to go?" I had missed seeing this one, but later, at the end of the fjord, I spotted a large dark shape curving up through the surface ahead. "There's a whale," I called out excitedly. "And another! I think they're pilot whales. I saw them in Scotland last year. See that domed forehead?"

As more whales surfaced, we could hear the deep whoosh of air as if down a broad pipe.

"Wow, don't they sound hollow." Then, I added, "I know pilot whales are supposed to be harmless, but I'm not sure about killer whales. Do you think they'd bite."

Geoff responded in an instant, "Oh, don't worry, you'd give it indigestion if one did! Anyway, just tell it you are too intelligent for it to eat. No whale would swallow that!"

We lost sight of the fjord behind us as we followed the cliffs, in shadow, around the headland. The sky looked on fire, but beneath the cliff was dark emerald, the cliff rising with patches of short vegetation and bare rock, ocher and purple. Despite the exertion, I felt cold and finally dared to admit it. "Geoff, do you fancy a brew-up?" Pulling ashore, I put on a long-sleeved thermal sweater under my fleece. With my anorak back over, my upper body was warmer, though the breeze still chilled through the fleece on my legs. I lit the stove, and then spread the map beside the chart. I wanted to compare the two, since sometimes the placenames differed.

"Across the fjord ahead, that end of land must be this narrow peninsula marked on the chart as Norðfjarðarhorn, *North-fjord horn or point,"* I guessed. "It's Barðsneshorn on the map, but look, it shows the bay as Norðfjarðarflói, *north-fjord-bay.* It's about seven miles to the far point." We studied the map together.

"Shall we head straight there?" It seemed the obvious thing to do, to cut across a single entrance rather than detour inland to where the fjord divided.

"Why not?" agreed Geoff. "Maybe we can get as far as there," he touched the map at a northeast-facing bay on the far side of the peninsula. "Does that say Sandvík? *Sandy Cove?* It looks like a beach on the map."

I felt joyful. We had reached Iceland and were truly on our way, carrying all we needed to be independent. The ocean was calm, the morning, which it had become by now, was spectacular.

The break, hot tea, and extra sweater had surely helped, but my legs would be warmer in the narrow space of my cockpit, sheltered from the wind. I was ready to get moving again,

Around the cliffs of Norðfjarðarhorn, the air was frantic with puffins and guillemots. Birds gathered in rafts on the water, cackling and grunting, grating, and shrieking. As our kayaks approached, they either dived or skittered away across the surface. There was a musty smell of guano, the undulating surface streaked with slicks of oily froth and floating feathers.

Seals lounged on the rocks, shrugging, and shifting, arching plump bodies to lift both head and tail simultaneously. Watchful and wary, they seemed reluctant to abandon rock for water in the

Seyðisfjörður

face of a questionable threat. A steep cliff towered high above the vast ramp of tawny scree which sloped from the water. Resounding thumps echoed crisply from swells hitting the shore, the chilly air somehow accentuating the clarity of each thud and crunch.

I was weary by the time we reached Sandvík. The sand curved away for a mile to the next headland, so it was a relief when Geoff suggested aiming for the nearest corner of beach. I chased a swell as it surged onto the steep grey sand, tumbled inelegantly from my tiny cockpit into the sucking backwash and wrestled my kayak clear. Then, we carried the kayaks one at a time to the edge of the scraggly grass.

It was four-thirty in the morning. Excited to reach our first camping spot, I saw to my delight we had the whole area to ourselves. The valley stretched away, narrowing to an end a little more than three miles to the southwest, mountains all around. Curving around to the next headland was the steep dark beach, the pounding breakers spreading white fans up and down the jet-black wet sand. We scaled the steep grey slopes and berms to the flat top of the bar.

Both of us were keen to explore. The map showed abandoned farms a half-mile inland. From where we stood, we could see only a single signal-orange building in that direction. Geoff suggested, "Let's go and check it out."

Hemmed in by and hidden behind the strand lay a series of lagoons and waterlogged islands. The islands spat clouds of shrieking terns at us as we skirted the pools and splashed through the shallows. Whimbrels and golden plovers called out in fluting whistles. At the pool's edge, feathers trembled as if trying to break free from the foamy waterline.

Far beyond our goal, the valley ended as a steep face, cutting a jagged skyline up to 2,500 feet above the valley floor. Snow plastered the back wall and lingered elsewhere in cracks and gullies, white scraffito patterns crisscrossing the lower crags. The steep slope leveled twice in steps, each edge throwing white rivulets of meltwater together into ragged gleaming ribbons running down the cliffs toward us. Eventually, the streams sank from view into the meadow below.

I ran my eyes along the skyline. The ridge held its elevation, with only minor dips between this side of the bay and the headland on the other. A tower of rock, dwarfed by distance, stood like a figure on the ridge. Somewhere overhead a snipe was drumming, each new dive a trembling crescendo.

Hiking easily across the sandy grassland of the valley floor, we approached the few ruined buildings, the stone and sod walls engulfed by grass. Each had become the repository for the tangled remains of its turf roof fallen in. Nearby, the brightly painted building stood square and tidy beneath its corrugated metal roof. This, according to the sign on its wall, was a shipwreck shelter.

"This is sturdily built!" said Geoff, impressed, as he swung the heavy door outward to find a second strong door, which opened inward, within the same frame. He sized it up with his carpenter's eye before stepping forward. Inside were provisions, a radio, bunks, and instructions.

"I read up about these rescue huts," I blurted, "They're all around the coast, in places where there's no help nearby. The fishing companies fund them, to help boat crews after shipwrecks."

Geoff seemed distracted. "I could live in a place like this," he murmured, peering out through the window thoughtfully. "Just look at that view." I joined him at the window and saw ribs of purple-grey rock, and broad streaks of grassland sloping from the mountains into the valley.

"Well, let's spend the summer here!" I joked.

Seyðisfjörður

Geoff looked tempted, a quizzical smile flashing across his face, as if for a moment trapped in a dream. He shrugged it off and said, "I feel peckish. Shall we head back?"

Now meandering, we stumbled over four eggs, in a shallow nest on the ground, and carried them back with us. Collecting sticks of driftwood, Geoff lit a fire, and we fried and ate the eggs.

Sat there on the gravelly ground by the fire, I began fingering the pebbles. Some felt smooth, each colored red, yellow, or green. Others looked like cartoon asteroids, pitted, or studded. Tufts of scrawny grass, and small wiry tangles of plants, clung between the pebbles. Farther along the strand, driftwood trees, silver-grey and bone-smooth, held twisted arms into the air.

Nearby, the rust-umber shell of a Second World War sea-mine lay like a giant blown egg. A drift of coarse sand reached from the beach, into the hole in one end and out from the smaller hole at the other. The beach held it in its grasp, gently tethering this former sea weapon, claiming it for the land, tugging it down little by little into the underworld. It was a single, now-harmless outer case. Yet, between 1940 and 1943, the British Navy laid a hundred thousand live sea-mines between Iceland and the Faeroe Islands, in wait for ships or submarines. Today this relic, its wartime vigil over, lay redundant and at peace.

The structure of a wrecked boat, half-buried in the beach farther along, spurred us to take a closer look. Only the rusted metal superstructure of the fishing vessel protruded from the sand, leaning at forty-five degrees landward. Sand completely buried the shoreward gunwale. We clambered and clunked all over the sturdy shell, the hard, crinkly patina of protective rust burnished by sand and sea.

Curiosity satisfied; we trudged back beneath a clear blue sky. My legs grew heavy. After thirty miles of paddling in daylight, I was ready to sleep, yet instead of the soft approach of dusk darkening into nighttime, the day was waking up and growing brighter. We unrolled our sleeping bags onto the sand and went to sleep.

Little flies, running across my face, woke me about four hours later. Swatted away, they came right back, keeping me from sleep. When I could take no more torment, I got up, lethargically poked the fire with a stick to stir up glowing embers from deep in the ash and fanned a flame. Balancing the already charred ends of driftwood across the fire, I filled a pan with water and balanced it over the heat. Stretching and yawning, I knew that coffee would help. With woodsmoke drifting into the air, sometimes onto me and sometimes away, I sat cradling my mug, too sleepy to avoid the smoke.

Once I had finished my coffee, I took to my feet and wandered to where the tide had exposed the rocks. There I collected a handful of glossy blue-black mussels. Cooked before breakfast, they made a salty, smoky, fishy, if gritty snack.

It took me time to get afloat, frustrated loading my kayak. When I could not squeeze everything in, I pulled it all back out, thought about it, and started afresh. Geoff waited patiently until I was ready to help carry the kayaks to the water, and we could aim across the bay.

The coast rose steeply in layers of alternating ramps of scree, and ramparts of cliff from which the scree had crumbled. In all, it towered two thousand feet or more. In places the scree had grown over with grass, or whatever arctic-alpine plants might grow up there. Other scree slopes were of bare rubble. In general, the mountains looked craggy, steep-stepped, and mostly green. Here and there, white slopes of snow reflected brilliantly in the sun, and pinnacles surmounted the sharp-cut ridges edging the skylines.

We headed across Reyðarfjörður, four miles of open water, toward Vattarnes, a low rock peninsula with a yellow lighthouse. I still felt weary, so even that short crossing seemed endless to me. It was worse for Geoff who had to match my slow paddling pace.

Seyðisfjörður

On reaching the other side and taking a break close to shore, Geoff seized the opportunity. "I'm going to catch a fish!" When he reeled down his line, all his confidence came to nothing.

"I don't think my line's long enough to reach the bottom," he observed. "Either that, or the sinker's not heavy enough for our drift." I looked up, surprised to see how quickly the breeze was pushing us sideways. If his lead weight were too light, it would drag along the surface. What would it take to hold the line down?

"Let's have lunch!" First, we needed fresh water. Beside the fjord, farther in, stood a cluster of houses, or a farm. We could easily paddle there, but tempted to land I suggested, "Why don't we walk back to get water?" That agreed, we scouted for the easiest place, near the north end of the point. Puffins whirred past with loud wingbeats. Gulls drifted, hanging in the air above us, shrieking.

We chose a rocky beach, the boulders splattered with bird droppings. Geoff scoured the tideline, throwing together chunks of driftwood for a fire. Having fetched water and set a pot over the edge of the flames, we stood and waited for the water to boil. Geoff's fire soon blazed high, thrusting out a dense plume of smoke that chased and clung to us, no matter to which side we stood. We circled, trying to keep close to the fire, to enjoy the heat yet avoid the smoke.

Between the boulders lay fishing floats, each a foot or so in diameter. I picked up a tough orange plastic one. Others of metal looked like galvanized steel or corroding aluminum. Geoff posed, cradling a football-sized float under each arm, and *put the shot* with them. They clacked on impact, bouncing from rock to rock down the shore. I was surprised how much they leaped. The floats, hard as rock, had no spring in them. I suppose the density of the plastic made them bounce so high. There must be thousands of fishing floats floating freely around the oceans of the world. Does anyone ever collect them to reuse?

Iceland by Kayak

Having cooked over the fire, and filled ourselves with coffee, we doused the flames and prepared to leave. Our route would cross Faskruðsfjörður to Hafnarnes.

"It's really difficult to talk about places you can't pronounce," I complained. "I can't even read half of them."

"How about bananas," Geoff suggested, quizzically.

"What?"

"Hafnarnes, half-narners. Half-bananas," Geoff explained the evolution. "Half-narners, slang for *half-crazy*."

I laughed. "Okay, let's head for half-narners." That would be an easier name to remember. I wondered how many other awkward names I could break down in the same way.

We reached the easily pronounced island, Andey, and the current carried us onward toward Stöðvarfjörður, over which name my tongue stumbled. High in the mountains across the fjord we could see a magnificent tower of rock catching the sunlight. This was Snæhoammstindur. I could translate *Snæ* to snow, and *tindur* to peak. Was the middle part from *hamar,* a mountain cliff? Did that reference the cliff tower? Place names seemed descriptive.

"This scenery is fantastic! I love all the ridges and towers and spiky bits. It's like something from Tolkien's *Lord of the Rings*!"

"Yes. You could imagine trolls and giants up there, rolling rocks off the top," said Geoff absently, then adding what was on his mind: "How far do you want to go today?"

"Let's aim for Kirkjoból," I suggested. "We could get food and water there."

According to *Iceland in a Nutshell,* the entrance to the smallest fjord on the east coast, Stöðvarfjörður, opens four miles beyond Hafnarnes. There is a settlement of 241 people, just a short way inside the fjord on the north shore. It is sometimes known by the name of the fjord, otherwise as Kirkjoból. To stop there, would not involve us in more than a three-mile detour.

Seyðisfjörður

Drawing close enough to see the village, we discussed where to land. "I think a little way outside," I suggested, "to give us some privacy."

Geoff disagreed. "We might as well land right over there. Look. It's an easy beach and it's closer to everything we need." I looked to where he pointed. There was a small hut at the beach, next to a river, and a patch of long grass.

"Okay," I deferred.

We carried the kayaks to the grass, and I opened my rear hatch to look for a change of clothes. To my dismay the hatch had leaked. When I pulled my clothes from their drybag, I saw damp patches where the drybag had leaked through the seam lying in the water. I was glad the clothes were no wetter.

I turned. People were gathering around us, with at least a dozen already quietly watching. What should I do? I was ready to change from my wet stuff, but I felt uncomfortable, shy. I regretted agreeing to land so close to habitation.

In silence, everyone watched intently to see what we would do. I looked to Geoff for a lead and caught his eye. He shrugged and continued to pull items one by one from his drybag, fussing, killing time.

Having hesitated for long enough, I wrapped a towel around my waist for modesty. Nobody seemed inhibited to stand and stare. Was it a purely British characteristic, I wondered, when itching to know what is going on, to pass by without breaking step, with no more than a casual glance sideways?

I felt better in jeans, t-shirt, and fleece sweater again, even if they were a little wet. I was proud of that blue fleece. I had bought it when my climbing friends at college introduced me to Helly Hanson's ground-breaking fiber-pile. Geoff preferred wool, even when paddling.

When our crowd, having seen what they could, drifted away, Geoff strung a line for our wet clothes. We left them dripping, while we carried our empty milk jugs in search of fresh water. The river looked too dirty, and at two in the morning, with nobody left in sight, we did not want to knock on doors. What had we been thinking? There had been a crowd of people to ask earlier. We had missed our opportunity.

I looked around at the cheerfully painted walls and roofs of the houses, which contrasted the subdued greens and greys of the surrounding landscape. Painted on the walls of one house were giant chess pieces, and on another, huge artist paintbrushes. Someone had painstakingly painted the walls of one to look like clay bricks, an incongruity in a land of earthquakes.

A car approached, scrunching unhurriedly down the dirt road with two girls inside. We waved it down and asked where we might get water. "Have you knocked on any doors?" the driver asked.

I pointed to my watch. "It's rather late, isn't it? I don't want to wake anyone up." She looked at me as if bewildered but took our containers through her window anyway. "I'll go to my sister's. Wait here."

She drove down the road a short way and stopped. We watched her disappear into the house. When she returned, with our bottles filled, she also handed us a plastic bag full of Danish pastries. "Wow! Thanks so much!"

"You're welcome," she smiled, and drove away. We returned to our kayaks, munching as we walked. To be sure we would not wake up with another crowd watching us, I pitched the tent. Geoff fired up the stove for a brew.

Seyðisfjörður

SANDVÍK.

Iceland by Kayak

SEA-MINE, SANDVÍK.

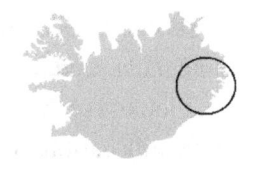

8
A Taste of Wind

I awoke to the sweet scent of meadow, the sound of terns crying and wavelets lapping the shore. Keen to see what I could hear, I hurried outside. There, hunger taking precedence over further curiosity, I realized we must shop before breakfast.

Returning from the small store, we laid out our purchases of onions, cabbage, bacon and cheese, bread rolls, orange juice, cereal, milk, and best of all, tea. We had been drinking instant coffee, in good supply, but had run out of tea. Unsurprisingly, that which we lacked we missed the most. Finally, resisting the temptation to eat the rolls before frying the bacon and onion, breakfast preparation could proceed at a leisurely pace.

Since we were camping at the edge of town, we lit the stove rather than a fire. On our single burner, we cooked breakfast before boiling water for the longed-for tea. But as I added the fresh milk, I watched in horror as congealed white lumps surfaced in my mug.

Iceland by Kayak

"Oh no, the milk's off," I groaned in dismay. But seeing its consistency, I looked more closely. "No, I think perhaps it's just some other kind of milk product."

I looked at the label. Icelandic made no sense to me. It looked like a milk carton, so what else might I expect? "I think it's a runny kind of yogurt," I guessed eventually.

"Bummer," said Geoff. "We'll have to make do with tea without milk."

On the water we backtracked, hugging the shore to the headland near the lighthouse before cutting straight across the fjord entrance. Bands of mist around us made the mountains appear to float in the sky.

Once past the next headland we started across another bay. I felt comfortable with my pace, but Geoff was faster than me. It irked me to be always just a little behind him, always at about the same distance.

As we paddled, the mist thickened around us, quietly enveloping everything. It obscured what had been a spectacular view of mountains ahead. The fog must have lain quite thin above the water, for despite the poor visibility, the sun still beat down on us. Feeling the sun more than the fog, Geoff announced, "I'm too hot in this jacket. Can we raft up for a minute? I'm going to take it off." I took the opportunity to do the same. "No doubt we'll go back home with lily-white bodies," Geoff grinned. "Are you ready to get going again?"

The sun was strong enough to steer by in the fog, so we did not need to pay attention to the compass. It seemed ironic that yesterday we crossed Reyðarfjörður, beside which is an old calcite quarry. That quarry was a source of Iceland spar from the 17th century until 1924. I have heard it said that early Icelanders sometimes used Iceland spar in fog, and on cloudy days, to tell the direction of the sun. They called the crystal *sólarsteinn*: sun stone; otherwise *silfurberg*: silver rock.

A Taste of Wind

Iceland spar both polarizes and causes double diffraction of light. If we viewed an object through the crystal, we would see two images caused by the double diffraction. Viewing sunlight through the stone, we should see two different intensities of polarized light. Rotating the stone changes the intensity of each, showing a greater or smaller difference in that light intensity. By rotating the crystal until the intensity of each is equal, it is possible to tell the direction of the light source, in this case that of the sun, even when cloud or fog hides it from sight.

If the layer of fog around us grew deeper, hiding the sun, we would have to resort to paddling by compass. Had we a crystal of Iceland spar, in theory we could continue to use the hidden sun.

Somewhere in the middle of the bay we passed rocks with waves washing over them, the swell gentle and regular. But when conditions began to change, the wind rising, we landed at the first opportunity to warm up. This was on a small island, Hafnarey, which had a sheltering brood of off-lying rocks.

Moments after landing, Geoff abandoned his kayak at the shore and hurried up the hillside. Puzzled as to why, I followed.

"Ducks," he called over his shoulder. He had never called me that before. Was he thinking of dinner ducks, I wondered? Was he *Hunter by name and hunter by nature*? Then, my foot broke through the springy turf into a hole. I tripped and fell. I now saw puffin burrows all around, little holes perforating the slope. There were also dozens of eider nests, each a cozy hollow on the ground, lined with thick cushions of pale grey-brown down. In each nestled eggs, beautiful avocado-green duck eggs. I was growing hungry. An egg would taste good, and this early in the season, the duck would lay a replacement for any missing egg.

Geoff on the other hand, more ambitious, was searching for an unwary duck. I left him to it and wandered along the cliff, watching the fulmars slide aloof on the air currents. Stumbling at another puffin burrow, my foot punched through the turf onto the puffin inside.

Icelanders eat puffins. They catch them by the thousands. They usually whisk them from the air near the cliff colonies using long-handled nets like lacrosse sticks. One puffin would scarcely satisfy two paddlers, so I tried my hand at catching a second, crouching low beneath a cliff edge on which puffins were landing by the score.

By the time I returned to the kayaks, Geoff lay stretched out on his back, with his hands behind his head, looking very relaxed.

"No luck, no duck," he announced with regret. "But we can make an omelet." He broke into an agreeable smile.

"Too bad about the duck," I said, quietly laying two puffins onto the grass beside him. His eyes lit up and he rolled onto his elbow to look.

"Good job!" he said. "Do you want to find a better place to cook and eat?"

We studied the chart together. "How about Djúpivogur?" a good distance, yet not too far.

The breeze cleared the mist from the water until we could see our heading as we paddled, and when the wind died again the mist stayed away.

A huge waterfall fell from the cliff to the north side of the fjord, across from Djúpivogur. It captured our attention as we drew closer. Eyes alert, Geoff pointed out, "Hey look, there's a ton of driftwood over there, on that beach." He looked to me for a positive reaction. "Do you want to stop there?"

"Sure, that looks great." We turned toward the shore and pulled up onto a rocky beach. There, along with the sea-heaped driftwood, lay the huge skull of a whale. The bone was dry and beginning to crumble, so it had been there for a long time.

Geoff began gathering armfuls of driftwood, piling them in a sheltered spot for a fire. By now I had learned that few things pleased him more than building fires on beaches. This one was terrific, warming us until we had a bed of glowing embers over which to roast the puffins. With mussels gathered from nearby, onion, puffin liver, and cabbage, plus a plump puffin each, we were in culinary heaven. The puffin flesh was tight like pigeon, and wonderfully kipper-flavored by the wood smoke. We sat back and stared across the fjord. "Yum, that was good!" I declared, the flavor lingering in my mouth.

"You bet," agreed Geoff.

The tide with us again, we passed offshore Djúpivogur, so far from shore that we did not even see the curving sandbars marked on the chart. Our visual target was a ridge of mountains on the far side of the next two fjords, at Eystrahorn, *East Horn*. We could clearly see the line of mountains with a roll of cloud hanging along the top, twenty-five miles away, but did not recognize its warning.

Eystrahorn was a prominent peak, marked as 2,478 feet high. While easy to identify on the sea chart, there was nothing of that name on the topographical map. Instead, the same mountain appeared as Hvalnesfjall, although the dark map symbol, showing steep rocky slopes, blackened part of the name, making it difficult to read.

The intended meaning of *Hvalnes,* a low point jutting south on the chart, was clearly *whale nose,* whether the nose had given the mountain its name or not.

As we paddled, Geoff remarked on my paddling pace. "Can you paddle faster?" he asked.

"Well, I could," I admitted, "but if do, I don't know if I'll have enough in reserve later, if I need it. I know I can keep going like this." I thought I knew my capability. Aware of Geoff's continued frustration, even if he was too much of a gentleman to push the point, I did increase my pace. Was this the time to find out how long I could keep it up?

We made comfortable progress with the help of the tide. When we reached land again, black sand beaches fringed the huge scree slopes that ran down from crumbling crags.

"Stop here, or press on?"

"Let's press on. This looks too bleak. We don't have any water left and there's a river farther down, just before the next headland, before the lighthouse."

The lighthouse, marked on the chart with a magenta-colored tadpole shape, was easy to spot on the otherwise white and pastel colored sea chart. The black line of the river showed clearly too. On the topographical map the river showed as a blue line, flanked by green, sandwiched between tightly packed brown contour lines. Beyond the point with the lighthouse ran a long curving sandbar, enclosing lagoons. There would be no value in going farther than the lighthouse tonight.

But as we continued, the wind sprang up against us. There was also what looked like a tide race out at sea, although it was debatable, from the way the waves were breaking, whether it was from current, stronger wind, or both. I could feel a current running here and with its assistance, we soon rounded the next corner of coast, pushing now against a quite strong wind.

Ahead, the mountains sloped up to an imposing wall of rock, capped by a roll of brilliant white cloud which obscured the summit ridge. This was the same cloud we saw earlier from afar. It flowed outward from the ridge and dropped toward the valley like a giant waterfall. Constantly replenished from above, unlike any waterfall it evaporated mysteriously on its way toward the valley below.

Dwarfed beneath the mountains lay a sickle blade of grey sand beach, its surface twisting into drifting waves of misty grey and rearing snakes of sand.

The closer we approached the mountain, the stronger the wind grew. It whipped across the water, one moment cold, the next hot, lifting stinging lines of spray and wrenching at my paddle. I struggled, anxious to both stay upright and keep moving forward. After each major gust I grimaced at Geoff, and he frowned back. He was not enjoying himself either. I tried to stay near him, without getting close enough for the waves to throw us together.

We aimed toward the very end, where the tip of the beach touched rock, seeking the place offering the most shelter. If we landed sooner, we would have no shelter at all, so we punched on, directly into the wind, until we eventually reached the weather shore. There, we ferry-glided across the wind to the most sheltered corner, swearing and laughing at having made it.

"That's stretching it as near to the limit of endurance as I want to go," admitted Geoff. "Sometimes I wasn't moving forward at all."

I laughed. "Some of those gusts were hot, like from a hair drier!"

"I know. They felt incredible. How much do you think it's blowing out there?"

"Strong. It must be gale force to raise all that spray."

"Best get these kayaks up a bit and unload them. Do you want to change first, or put up the tent?" We decided to put the tent up, so we could heat water for a brew while we changed.

We lifted the kayaks well clear of the water to unload before carrying everything, including the empty kayaks, far beyond the beach. Next, we searched around for somewhere for the tent, any clear flat area in less wind.

There was a huge empty mud-pan, obviously once water-filled, now dry and whipped into mini dust storms. Beyond that spread a grassy slope leading up to the mountain. My walking pace fell to half its usual vigor against the whistling wind. I felt exhausted. It was desolate here, more like desert than anywhere I had ever been, and there was no protected spot.

"Well, if there's nowhere sheltered, here seems as good as anywhere, doesn't it?"

"Yeah, let's get that tent up."

It took both of us to wrestle the tent onto its frame. It strained, like a living beast, pulling and struggling against our attempts. We pinned it down with rocks all around, but even then, the fabric bulged and slammed in protest.

Next, we needed water, so I took an empty container from my kayak and walked downwind into the broad open valley, to find the river that lay hidden by the stony terrain. Buff-colored balls of down, the size of melons, bounded and tumbled past: the lining from eider nests, while the strange falling cloud reflected an eerie light into the valley.

I paused for a moment, balancing with difficulty on the top of a boulder, while the wind pushed and shoved. All around me stretched gravel banks and mounds of pale boulders, the valley stripped of vegetation if any ever grew there. When the river ran high, all these rounded rocks must trundle downstream. Despite the desolation, or more likely because of it, I appreciated the strange beauty.

My eyes watered in the wind, and I brushed them, cursing at the sting of salt from the dried crust on my face and eyelashes. My eyes were just clearing again when I thought I saw a movement. I looked more carefully. Although it might have been a bird, I had the impression a boulder had moved.

I was about to turn away when I caught another shift of shadow. There it was again, a reindeer! I whistled quietly through my teeth in glee, watching the pale phantom drift across the rocks. Suddenly, I saw others. I counted six, then more. There were ten, eleven, a dozen at least. They were little more than one hundred yards away, so either I did not bother them at this distance, or they were unaware of my presence.

A Taste of Wind

I moved stealthily toward them, aware that any sound or scent would carry away on the wind. Their pale bodies floated gently and effortlessly as if the boulders never existed. I followed for a full ten minutes, marveling at their almost perfect camouflage. A motionless reindeer was almost invisible at this range, yet they were not tiny animals. I recalled having read somewhere how reindeer was an introduced species, brought from Norway in the eighteenth century. I had not expected to see any, so I felt privileged.

I retraced my steps down the valley, filled my container with cold clear water and pushed myself back against the wind to the tent. Every step of the way demanded a ridiculous amount of effort. In the relative shelter of the tent, the petrol stove roared and stuttered. The flame burned blue, then billowed yellow, flaring from beneath the pot with every slam of the wind and coughing black smoke. I shut my eyes and groaned. "Wake me up next week," I muttered to Geoff. But, although weary, I was not about to go to sleep before supper.

HVALNES. AUTHOR FETCHING FRESH WATER.

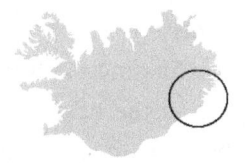

9
Pinned

I awoke sweating in my sleeping bag, buffeted by the tent slamming around me. The sun-soaked orange nylon whipped back and forth, the sides beating inward. So far, the rocks still held the tent in place. Otherwise, with no sewn-in groundsheet, the whole shelter would have blown away leaving us in the open.

As I crawled out to view the day, sand filled my eyes. The wind howled. Drifts of sand and dust flew like shrouds of mist. At sea, the waves breaking near shore disintegrated into dazzling plumes of spray that streamed back across the bay.

I zipped the tent closed behind me. From inside I heard muffled groans as Geoff stretched. He sounded comfortable. "Oh, what a wonderful morning!" he called. Soon his cheerful face grinned from the entrance.

To get a better overview of the conditions, we set off walking in the direction of the lighthouse: a squat yellow concrete tower on the low rock ledge at the end of whale point. I pushed zombie-like against the wind, my feet not always landing quite where I aimed them. My jacket held me in a tight embrace and the excess fabric in my jeans shuddered behind me, blown back by the wind. Breathtaking, it was tiring even to stand in one place. "It's bree-

Iceland by Kayak

ee-eezy!" I shouted, drawing out the word. The wind made my ears ring.

Finding the lighthouse door locked, we stood outside to look across the next bay. The shore was a long grey bar of curving beach with a lagoon spread wide behind it, a little reminiscent of Chesil Beach in the South of England. A ridge of mountains jutted out like a steep dark wall, fifteen miles away, at the far end of the bay.

Despite the frenzy of whitecaps, the sea did not look particularly wild overall, considering how strongly the wind blew just here. I knew this was deceptive. Yesterday, I struggled to make progress in my kayak, challenged to stay upright. Today, it was difficult enough to keep my feet on the ground in the wind. I was not keen to repeat yesterday's experience, not even if conditions were at their worst only at the headland.

"Doesn't look too bad really," I teased, tongue-in-cheek. "Not too much swell. We could easily zip over to that next headland and stop for an early lunch." In truth, for the effort it would take to round this headland, leaving today would be pointless.

We circled the lighthouse to look inland. The cloud-topped mountains appeared rugged. Beneath the closest mountain, on the west side near the shore of the lagoon, stood a small farm. It was quite close, less than a mile.

"Let's take a gander," Geoff encouraged. A rough track led from the lighthouse toward the farm. At the farm we reached the start of a gravel road. According to the map this led to the Iceland ring road, four miles away, meeting it where the ring road climbed to cross the mountains, avoiding the headland.

Close to the farm, a little girl came out to meet us, questioning us in Icelandic. We could not understand her, and she looked puzzled, disappointed when we did not reply. Nearby, a young man worked under a car. If he noticed us, he did not show it.

Pinned

Here, by the farm, there was hardly any wind, so we drifted away to find a place to sit in the sun, whereupon a woman came from the house. She spoke a little English, so we asked, "Do you have any bread we could buy from you?"

She invited us inside. My ears, already used to the outdoors, felt odd hearing the close echoey sounds in the calm of a small kitchen, as if I had just removed earplugs. I sighed contentedly, looking out through a window at the view, with no wind to pull tears from my eyes.

"Please, sit." She began to prepare toast, and salad, fruit, and cream, milk, and coffee, setting the table in front of us and encouraging us to eat.

"This tastes so good," I enthused, thanking her as I ate.

Geoff quietly explained, "We're paddling kayaks along the coast, starting from Seyðisfjörður. We want to try to make it all the way around Iceland. We got here last night, and the last bit was windy, right at the headland here."

"Yes," she agreed, "it's an interesting headland. The Gulf Stream and the Arctic current meet in the middle of the bay just past Hvalnes, this headland. My brother is a pilot for ships entering Höfn, the town just down the coast. He knows all about the currents."

Would that be the Irminger Current from the Gulf Stream, meeting the East Iceland offshoot of the East Greenland current? I had no time to ask before the door opened, and a girl entered the kitchen with a calf.

"Wow," I looked at Geoff, "is that a reindeer?"

"Yes," the woman answered, "they come down from the mountains when the weather is very cold." The calf was tiny, with long thin legs. The kitchen floor was too slippery for its hoofs, so it skidded and slipped, its limbs out of kilter when it moved. The girl rushed to support it, hugging it lovingly against her. I presumed it was her pet.

Iceland by Kayak

When the girl took her calf away, I took the opportunity to ask, "Is there a shop anywhere near here?"

"No, but if you like, you can come back in the morning and phone the postman. He can bring food from Höfn. You can get fresh milk from the farm here in the morning too."

Revived by the food, we thanked her and left, briskly retracing our path toward the lighthouse. Hitting the wind once again drained my whole body of energy. "I'm knackered," I admitted. Geoff must have been too, for on reaching the tent, he stretched out on the ground nearby to rest. For a time, I did the same, until the relentless wind bothered me too much. To get out of the wind, I crawled into the tent to sleep, but it was too hot in there. When I overheated and Geoff got too cold, we switched, and later switched again.

With the tent pummeling, there was only room for one of us at a time inside, lying right along the middle, otherwise the fabric shoved at us. Outside, when we tried to rest on the ground, sand and debris bombarded us. Eventually, both awake enough to eat, we cooked a curry which we washed down with tea. The wind showed no sign of abating, and the tent strained alarmingly, so we piled more rocks to better secure it before crawling inside to study the map.

"If we left here early in the morning, do you think we could make it to Höfn before the shops shut in the evening?" Geoff wondered.

"How far is that do you think?"

"About twenty-five miles, thirty? That should be do-able, shouldn't it?"

The constant daylight meant we could easily paddle at any time of day or night. Still, I was doubtful we could leave. The wind seemed to have settled in to stay.

Geoff dreamed of a more optimistic scenario. "I'll get up early tomorrow in the blazing hot sun, panda across to the river in my shorts, have a really good wash and be dried by the sun in seconds."

"More likely freeze-dried," I muttered as a particularly violent gust flung gravel, rattling, against the tent.

The following day began sunny and cold, with the wind still thrashing and ballooning the tent. The cloud pouring from the mountain continued to mesmerize me. It was just like a massive waterfall dropping in a fast-falling, billowing curtain from the mountain ridge. It was magical how something of such apparent substance could dissolve into clear air as it reached a lower, warmer, elevation.

The sea, still flattened by gusts, appeared a little less white than yesterday. We walked around the headland to get a fuller picture.

"We could probably get around the headland past the lighthouse at least," Geoff judged at last, to my dismay.

"It would be really demanding," I protested. I wanted the wind to stop before we left.

"How about leaving later today, to at least make a bit of progress?"

We agreed in principle to a later departure and after breakfast I walked down to the river to bathe. It was worth the pain to feel clean and fresh, even though the numbing water with chill wind felt excruciating. It was an opportunity to wash clothes which I expected to dry quickly. Instead, the wind took them away, unless secured with large rocks beneath which nothing dried.

The rear hatch of my kayak had been leaking, so Geoff helped me carry the kayak to the tent where I set about altering it. The hatch sat in a deck recess with no drain-away. The hatch lid, the disk of plywood with its waterproof cover, sealed around the hatch rim with bungee, like a spray deck. I realized how the lower edge of the fabric must sit all day in water pooled in the recess. Any change in pressure would suck water through the tiniest of imperfections in the seal into the rear compartment.

As an added layer of defense, I made a waterproof nylon sleeve and glued it around the hatch opening. By gathering, folding, and tying the neck of the sleeve, I would create an extra seal. The original hatch lid would fit over the top.

By the time I finished I was tired of the wind lashing me, and of the spatter of sand, grit, and dust, so I sat with Geoff in the tent. That offered only partial respite. The tent proved to be a restless shapeshifter. We sat there chatting until Geoff stopped abruptly, mid-sentence, to announce he was hungry.

"I'm famished too," I admitted.

"I'd settle for roast beef," Geoff began. "Roast beef, Yorkshire pudding, sprouts, roast potatoes, and carrots. With gravy of course. That should do for starters. What do you think of that?"

"Well, I'd be happy with a huge steak, with fried onions, tomatoes, and salad..." I trailed off as the mental image grew vivid, becoming too absorbed by what I had conjured up.

"Black Forest gateau or blackberry cheesecake?" asked Geoff, "or a bit of each?" Hunger always brings a conversation back to the baseline, food. A good enough reason for a cave man or woman to draw images of reindeer, or bulls, with hunters in pursuit. Even without words, pictures could convey the message perfectly well. "I'm going out to get some steak," here pointing at the bull picture. Next, pointing at the drawing of two hunters, "Will you join me?" Or "This one is you... bring back one of these."

"You know, there must be plenty of food we can find out there," Geoff said, "I'm just going outside," He paused, adding in a deliberately deep serious voice, "I may be some little while."

Recognizing that he was misquoting Captain Oates' last exit words, before his demise on Scott's historic Antarctic expedition

Pinned

to the South Pole, I fell in with the game. "But there's no need to go! We've got plenty of food!"

"But what food have we got?"

"We've got Scott's porridge, Oates." I recalled the kilted figure, putting the shot, pictured on the box of that product. In truth we had little else left besides porridge oats, and rice and onions. Geoff's descriptions of food had made me realize that I craved meat.

"Well, I'm going hunting," said Geoff. With a final grin, he zipped the tent behind him, leaving me to my thoughts.

I was sitting outside when Geoff returned. He looked proud of himself, clutching a dead duck. He laid it on the grass for me to admire, before announcing, "I'll go down to the sea to pluck it."

I set to work preparing a fire. A fire would be impossible without shelter from the wind, so I rolled boulders together and scooped out a hollow between them. Splintering a piece of weak driftwood, for kindling, I managed to coax a flame onto it. Once alight, the fire, fanned to fury by the wind, ate rapidly through my small pile of driftwood. I ran to find more, half-expecting the fire to blow away in my absence.

Geoff's duck, dismembered, just fit into our largest pot. We added sliced onion, herbs, and a little curry powder. Geoff pulled the rim of a rusted oil drum from the grass and used it to help balance the pot over the flames. Focused on feeding the fire, I heard a loud thud behind me and reacted just in time to fend off my kayak. This, positioned as a partial windbreak, had rolled in the wind. The bags it had pinned down to dry flew away.

As I leaped to grab them, I heard a clatter. A strong gust had blown the pot lid off, together with the stone that I had placed on top to hold it. I retrieved the lid, only to see a red drybag take to the air. It bounded and rolled across the sand.

Determined not to lose any gear, I sprinted after it for a hundred yards or so until it hit the water, where it washed back into the shallows on a breaker. I waded in, captured it, and returned, swearing at the wind, my jeans soaked in saltwater.

Geoff saw the funny side of it all, and we were soon both laughing. I changed into my blue *Damart* thermal leggings, long underwear, and spread my jeans under rocks to dry.

When the duck had cooked for long enough, we put on a pot of rice to boil. "I'm really looking forward to this!" said Geoff, rubbing his hands together in anticipation. He tended the pots while I stepped aside to pee.

I disliked peeing in such weather. With no shelter, it made scant difference whether I faced the wind or turned my back. The gusts would buffet around me, flinging a spray of pee in all directions. I hated it. "Confucius says, he who pees in the wind gets his own back," I muttered to myself, swearing when the wind fulfilled that prophecy. I was in full flow, completely absorbed trying to stay dry, when a red Jeep came trundling toward me. I turned my back, embarrassed to be standing in the open wearing thermal underwear, peeing. But where else can you pee in such a wide-open landscape? The Jeep pulled up beside the tent and a man climbed out, followed by a youth.

"My name is Elias Jonsson," he announced in perfect English offering his hand for me to shake. "Doctor Livingstone, I presume," ran through my head. Instead, I said, "Nigel. Nigel Foster," and turning, introduced Geoff. Geoff stood beside me, spoon in hand. After introducing his son, Elias explained how he himself was a newspaper reporter from Höfn.

"I read an article in the newspaper about your journey and decided to try to find you," he explained. We talked a little longer, and then he looked down at the spoon in Geoff's hand and asked curiously, "What are you cooking?"

Geoff and I looked at each other self-consciously. We had learned that the eider duck, although quite common, is a protected species in Iceland because of the economic importance of its

Pinned

down, collected from the nests. Could we tell a newspaper reporter that we had caught a duck? Unwise.

"Oh, it's just a pudding," Geoff said, and adroitly steered the conversation toward the kayaks, pointing out the hatches, pumps, and compasses. There was a chart on the deck, under the bungees. Elias had been kayaking before, so he could imagine the coast from our perspective. He pulled out the chart and pointed out places between here and Vík where it might be possible to land or find shelter, and which sections were more awkward. Then, not easily thrown off his earlier scent, he became concerned about our food again.

"Will it be all right," he asked, "won't it burn? What are you cooking exactly?" he asked again.

Geoff explained casually, "Oh, it's just flour, fat, a little water with sultanas added. We call it pudding. We cook it for a long time. It'll be fine." I looked at the two pots and wondered how effective the deception had been.

We asked Elias about killer whales, something I was still anxious about, having no definitive answer from anywhere as to whether they could be a threat to kayakers. Elias said he thought not, if we saw just one or two. However, he added that they could be dangerous in packs. "You should be okay though. You should only see one or two, if you see any at all. They come across from Africa for the herring, which arrive in August, and rarely go farther north than Reykjanes."

Now we asked about his Willys Jeep. He pointed out that the air intake for the engine was inside the cab, ensuring supply when he made river crossings. "Those are huge tires!" admired Geoff.

"Yes, the wide tires are best for sand, and there's plenty of that on the south coast. They are not so good for this," Elias threw out his arm in gesture toward the boulder-strewn terrain. "It was not worth changing the tires just for today. But what about your food? I am worried. Will it not burn?"

Iceland by Kayak

We were too deeply into the fraud to retreat. It would be a simple courtesy to offer food to our guests, although challenging to explain how our steamed sultana pudding had miraculously changed into a duck. Thankfully, Elias contented himself by taking photographs. Finally, with a formal nod to each of us, he insisted, "Contact me when you reach Höfn. Here," he scribbled on a piece of paper, "my address."

With a parting, "Goodbye, and good luck," the father and son climbed back into the Jeep. After a couple of attempts, Elias roused the engine and they eased away carefully on their giant tires.

Geoff and I grinned at each other. The unexpected meeting had cheered me up enormously, and clearly Geoff too.

"Alright, how about some grub?" Geoff asked.

I lifted the lid to check our dinner had not dried up and burned. It had not.

"May I serve you some pudding?" We both laughed, digging enthusiastically into the tender succulent meat with the predictably soggy rice. "Geoff, thank you! This was well worth waiting for."

It was tiring to shelter in the tent, while it struggled to free itself from its moorings. It was less stressful to wander around. I began to check out the small plants growing nearby. There were tiny saxifrage flowers, of yellow, and of cerise. Aglow against the ground, a compact green puffy cushion of moss campion, smothered in tiny pink blossoms, looked incongruously tidy. I could only appreciate the sweet fragrance of its flowers, albeit intense, when I brought my nose close to the ground.

There was a small white campion, and a ground-hugging thyme that weaved a living web across the sand. Otherwise, short thin dry grass, mosses, and lichens completed the ground cover.

Sitting tight on nests tucked into hollows on the ground, eider ducks relied on their camouflage to avoid detection. When they flew from under my feet, they left behind a clutch of olive or khaki eggs, or sometimes tiny ducklings, nestled deep in the soft down lining of the nests. The ducks had chosen the most sheltered spots. Even so, it was amazing to see how well the nest lining stayed in place, despite the wind.

An oystercatcher, startled from its nest, joined its mate circling, protesting loudly and shrilly. Their smart black and white plumage, and prominent carrot bills, dazzled in the sun as they sped around.

Next morning, not yet fully awake, I saw the usual tiny looping caterpillars which had plagued us ever since we set the tent here. Newly joining them, already everywhere in the tent, were large caterpillars of a different kind. Moments later, my skin cringed into goose bumps when something crawled heavily across my bare back. I was instantly very awake, horrified.

It was clearly a downside to leaving the inner tent in England and carrying just the rain fly. Anything might crawl inside the tent, and over us, and often did. Despite the caterpillars, the heat, and the relentless shaking of the tent, we stayed in our shelter until quite late.

Eventually, we decided to revisit the farm to buy milk, and to phone the postman for food, so after a brew we started walking.

"It's just like walking up a steep hill," Geoff remarked as we pushed at the wind. In stunning contrast, once around the corner and close to the farm again, we met blissful calm, as before.

Shown into the living room this time, we saw stuffed birds, mounted for display, and a tiny live bird with a broken wing recuperating in a cardboard box. After a few minutes, the farmer's wife called us into the kitchen where she had prepared a table for us with toast and jam, cake, milk, and coffee. I felt embarrassed that she might feel obliged to feed us.

Geoff spotted a broken chair. "Would you like me to mend it for you?" he offered, trying to describe that he needed glue. His attempts eventually foundered with nothing resolved. Without glue he was unable to make the repair.

When we insisted that our half-gallon container held enough milk for our needs, the farmer's wife refused to take payment for what she considered such a small amount.

"We would like to call the postman, to order food," Geoff asked at last. "Is that possible?"

She looked at him. "Not today. The postman never calls on a Thursday. And tomorrow is a holiday, so there will be no shops open again until Monday. If you are still here, you may come here again around four o'clock if you like."

Back at our tent we finished the remains of the duck, after which, weary of the sound of the tent constantly hammering, I went for a walk, dropping into the river valley. As I headed up the valley, I spotted a small herd of reindeer again, ghosting across the desert of grey lichen-covered rocks. Returning, I watched as with every gust, the sea broke into silvery curtains that raced down the bay, rainbows playing in the spray.

At the headland, the conditions still looked unfavorable. I wondered if we would ever manage to leave this spot. Around the tent, the wind still carried drifts of sand and dust in great gauzy gusts. Tired of all the grit and the buffeting, I carried my sleeping bag farther up the hill and tucked myself down into a hollow. There, where it was a little more sheltered, and quieter, I was able to sleep in peace. But I awoke with a start.

Pinned

AUTHOR RETURNING WITH WATER, HVALNES.

Iceland by Kayak

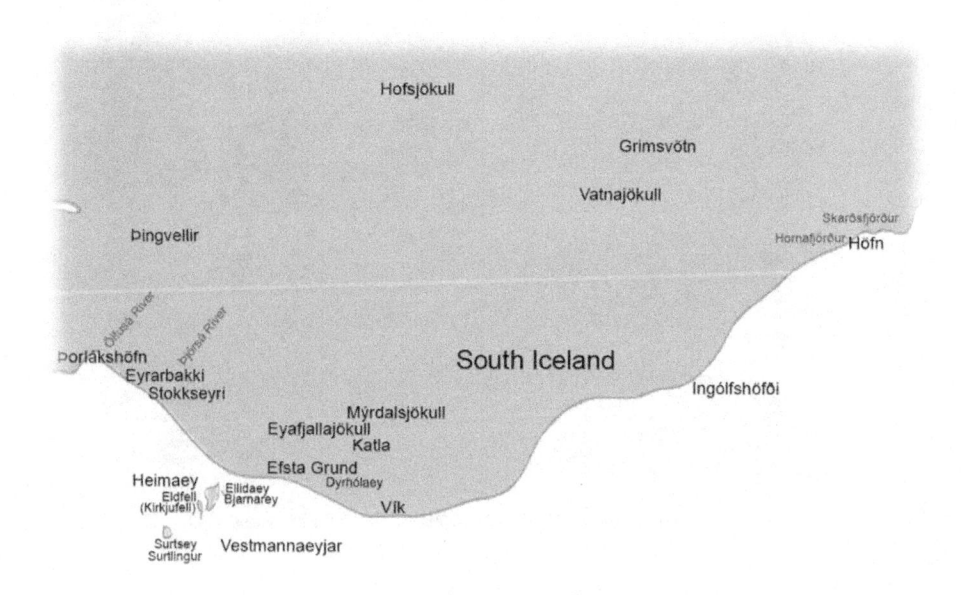

MAP 5. SOUTH COAST OF ICELAND.

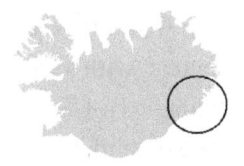

10
Höfn

It was just after three in the morning when I awoke and realized the wind had dropped. The sun was about to appear from behind the mountain and it was a beautiful morning. "The wind's gone! Yippee!" I shouted and dashed down the hill to start making coffee, and to waken Geoff. We were free to leave.

Launching with enthusiasm, we easily rounded the headland past the lighthouse. Ahead, two long curving sand spits defined Lónsvík on the chart, not a single bar as it appeared seen from the lighthouse. Each spit cradled a lagoon behind, leaving an exit at its western end. At the far end of the curving sand lay our next immediate goal, a block of mountains which, in the clear air, appeared crisp and deceptively closer than its fifteen miles.

About halfway there, near the island Vígur, we ran into muddy water. There was a well-defined edge between emerald water on one side, and muddy on the next. The less dense freshwater outflow from the lagoon pushed out to sea above the denser saltwater. The effect was visually dramatic and mesmerizing, a little like the bubbling edges of cloud formations. The next edge, crossing back onto the clear Atlantic, came not far after.

Iceland by Kayak

Closer to the mountains, which rose steeply to two thousand feet, we could see bee-hive shaped buildings on a lower headland beyond, with a white sphere and a couple of radar reflectors. These belong to a US radar station that has dominated this ness, or low nose of land, Stokksnes, since 1951. Its purpose until recently was to detect and track Soviet long-range heavy bombers and maritime reconnaissance platforms.

Soviet military planes routinely rounded Norway's North Cape on their way toward the Atlantic, passing through the gaps between Greenland, Iceland, and United Kingdom. These Soviet missions mostly probed the United States' air defense along the North Atlantic and the Caribbean. It was a NATO priority to show they always watched the gaps and spotted the planes.

To do this, in 1951, the United States Air Force installed radars at the four corners of Iceland: at Stokksnes here in the southeast, at Reykjanes in the southwest, at Straumnes in the northwest, and at Langanes in the northeast. The northern stations ceased operation in 1961.

Whenever the radar here at Stokksnes detected Soviet planes, the US Navy air base at Keflavík sent planes to intercept and shadow them. Technically now, in 1977, the radars here are non-operational, kept ready with supplies flown from the USA via Höfn, twice a week, via Keflavík.

As we followed the curve of the beach deeper into the bay the mountains obscured the buildings. Closer, they revealed them again, until we could more clearly see the telltale white golf ball on top of a building, and a group of huts. Nearer still, we recognized the American flag flying on a flagstaff.

At Geoff's suggestion we decided to stop, to see if we could get coffee. "Americans always have coffee," he assured me, confidently.

Assessing the surf, too much for a comfortable beach landing, we paddled past the American base. A little farther along, around the low cliffs of the headland, we probed behind a small rocky islet. There hid the gem of a tiny beach, in a gully between rocky points, sheltered by the outcrop.

We lifted the kayaks as far as we could and stood watching the waves clattering the stones in the narrow chasm, uncertain how far the water might reach. What if a rogue wave licked the cliff at the back? Geoff tethered the kayaks, just in case, before we scrambled up the back cliff from the gully and turned toward the radar station. Ambling along the open ground with no real plan of where to go, we saw the steep scree slopes and crags of the mountains we had just passed, a mile-and-a-half beyond the buildings.

Reaching the base, and with nothing to lose, we opened a door to one of the first buildings. Drawn by the clacking of a typewriter coming from a small office to one side, we stuck our heads around the door and asked, "Hi! Any chance of a cup of coffee around here?"

The man looked up from his typing and, without questioning our bedraggled appearance or our wet clothing, directed us to the next building. As we left, we crossed paths with two men, smartly dressed in military uniform, strolling toward us deep in conversation. One broke his stride to look at us questioningly.

"Coffee?" I asked. He nodded his head toward the door behind him. "It's free." He stared after us for a long moment, hesitant, before resuming his conversation, walking away.

As we crossed the threshold, all faces turned inquiringly. We stopped abruptly. "Is it okay for us to come in, wet like this, for coffee?"

"Sure, no problem," came a drawling reply, in an American accent. We made ourselves at home beside the coffee counter. Dripping saltwater onto the red carpet of the *NCO's Tiny Canteen*, we savored, and refilled, our mugs of rich strong dark coffee.

Iceland by Kayak

Nobody paid us any further attention, so when finished, we stacked our empty mugs beside the coffee machine. Calling, "Thanks," as we closed the door behind us, we sauntered back toward the cliff. When we dropped from sight into the gully, I was relieved to see the waves had not reached the kayaks.

Preparing to launch, I wondered, "Do you think we were supposed to be able to just walk in and have coffee like that? Isn't it a military site?"

"Well, they didn't seem to mind, did they?" Geoff replied rhetorically with a smug smile. We sealed our spray decks and shuffled ourselves like seals down the steep beach into the waves.

From the low cliffs of Stokksnes, we enjoyed a fast if wet ride, surfing open water waves for five miles along the sandbar, as far as the Höfn entrance. This single gap in the sandbank, teeming with seals and a porpoise, opened to extensive lagoons, and the harbor town of Höfn. Captured by the incoming tide, and drifting fast, we relaxed, letting the current carry us into the sheltered water.

Daydreaming and drifting, we ran aground abruptly mid-channel near a flock of cooing eider ducks. We sat there lazily and looked around us. The sky gazed up from the sheet of calm water. The shores, low green banks, melted into the distance. Mudbanks gleamed among clumps of floating seaweed, and here and there sat tiny, tufty, green pads of islands. There were buildings too, on the higher spots, but no obvious place to stop.

"Which way do you think we should go from here?"

"I've no idea."

"How about... let's try around there, to the left of that island." Close beside what looked like a small factory, we slid ashore onto what proved to be gooey mud.

Höfn, which means *harbor, or port,* is a town built on islands, extending south across the lagoon toward its narrow entrance. The town divides a large lagoon into two parts, east and west. With

Höfn

few other harbors on Iceland's south coast, Höfn is an important fishing port.

Someone walking past paused to watch, so we quickly called out, "We're looking for our friend, Elias Jonsson... could you help us find him please?" That he did, and soon Elias strode toward us welcoming. He surveyed our kayaks and looked us up and down.

"Bring all your wet things," he decided, "my wife will wash them, and they'll dry quickly in this wind." It was a wonderful offer, and we hoped his wife would not mind his generosity.

As we began to empty our kayaks for things we would bring, Elias spotted the radio in my kayak. He was curious. It was a Mayday radio telephone, a rectangular grey box with a massive red conical aerial. It was about four times the size and weight of a building brick.

This was the smallest portable, and waterproof, distress radio I could find when I was preparing safety equipment. It only just fit behind my seat, and then only because I had set the rear bulkhead far enough behind the cockpit to leave room for the pump. To access the radio, I would have to exit the kayak, so I doubted I could use it at sea. Designed for marine use, for calling from a life raft, it was impractical for someone swimming.

Elias picked it up and looked at it in wonder. On impulse, he turned it on and tried to call the harbormaster, just across the water from where we stood.

"This is no good," he announced, dismissively. "The signal is too weak even at this range. It would be better to shout." And he set it back down with a look of disdain. Geoff and I looked at each other as if rebuked. If wrecked, or stranded on an inaccessible shore, this was to have been our lifesaver.

Elias led us to where he lived, just a short walk from where we had landed. He introduced us to his wife, a small vivacious woman, and while she took care of our soft garments, we rinsed our waterproofs and hung them to dry. Despite it being National

Day, a public holiday celebrating Iceland's independence, there were two shops open, so we were able to buy the provisions we needed.

"Now coffee," announced Elias. Geoff rubbed his hands together, and we both smiled. *Coffee* represented much more than the excellent beverage. The spread was an open buffet including prawn pie, chocolate sponge, another chocolate dessert, and mini donuts, among other things. It was a long time since we had last eaten, and we ate ravenously.

Elias was a good friend of the Höfn policeman, so he took us to the police station to introduce us. He left us with him, saying he would find us later. The policeman watched Elias leave and then turned to us abruptly. "Okay guys! Down to the cells."

"You're locking us up?" pleaded Geoff in mock protest.

"Maybe he learned about the duck," I mouthed to Geoff as we descended to the cells.

"In you go!" The command from behind came sharp and serious. I crowded after Geoff into a cell. "Now you see the inside," he paused while the barred door slammed shut, "of a police cell, in Iceland." With one hand on the bars, he waved his finger threateningly.

There was a cold finality about his authority but, after a long moment of tension, he broke into a smile and continued, "There's a hot shower down here. Find me when you're clean and I'll take you for a ride in the police car."

That car turned out to be a huge Chevrolet, like a large souped-up version of a Range Rover. It was a bigger car than any I had seen on an English road. It offered a comfortable ride, smoothing out the bumps and potholes. As he drove, he explained, "They built Höfn on these islands in the last eighty years, joining the smaller islands with causeways, dumping rocks and shingle to fill in. The fishing fleet…" we could see the quite large wooden trawlers in the docks, "they are mostly trawling for lobsters at this

Höfn

time of year. They also catch…" he paused a moment, to glance sideways at me before finishing, "cod."

Of course, I caught the meaning behind his glance, and the slow deliberate way he had emphasized the word *cod*. Britain had just a year ago ended its third fishing war with Iceland. The sometimes-dangerous conflict was in response to Iceland's establishment of an exclusive two-hundred-mile fishing zone all around the island. The audacious extension of Iceland's territorial fishing zone threatened the livelihoods of all the foreign fishermen who had habitually fished there. Predictably, the UK bitterly opposed the idea. But the expansion had a precedent.

The first of the cod wars began in the 1950's when Iceland extended its four-mile limit to twelve miles. The second, when Iceland increased that limit to fifty miles, resolved in 1961. The third; the most recent cod war, began when Iceland extended the fishing exclusion zone to its current two hundred miles.

During the conflict, any British trawler caught fishing within two hundred miles of Iceland risked having its trawls cut by an Icelandic coastguard vessel. British trawlers refused to abide by the new limits and often suffered that loss.

Called to defend the trawlers, British naval frigates rammed the Icelandic coastguard boats, and the conflict turned quite nasty. If not for the cold war with Russia, with NATO's reliance on American bases in Iceland, the conflict might have escalated further. Britain eventually acceded, just a year ago. Those in the British fishing industry were furious. The loss of access to the Iceland fisheries devastated communities in Hull, Grimsby and in Scottish ports, where fifteen hundred or more boat crewmembers, plus thousands of shore-based workers, lost their jobs.

As unpopular as the decision was to the fishing industry, the British government had other fish to fry. The loss of those Icelandic fisheries would have little overall impact on the British economy. I nodded my head and smiled. My sea charts, printed in

1976, marked only the earlier twelve-mile limit, in magenta. The accompanying note read: *Limit of Iceland Fishing Zone resulting from the Exchange of Notes between Iceland and UK Governments of March 11, 1961.* "Cod, yes, cod," I acknowledged.

The policeman continued his explanation, referring to the fishing boats, "In August, the same boats are used for herring." I envisioned packs of killer whales competing for the herring, and was ready with questions, but he was in full flow.

"There is a small part of Höfn where refugees from Heimaey still live. Everyone evacuated the island during the eruption. Most people returned afterward, but not everyone had a home left. There were so many houses burned down or buried," he explained, "not everyone wanted to go back."

That eruption took place in 1973, four years ago. I remembered the images of molten lava, running like a river of flames into the town from Kirkjufell, while defenders sprayed the fiery rock with sea water. Initially named for the church location beside the first part of the eruption, the name of the new volcano, Kirkjufell, later became Eldfell.

We pulled aside at the summit of the road, to stand in the cool air at a viewpoint, overlooking the lagoon. Immediately below us stretched Skarðsfjörður, its ocean shores a low tongue of land reaching away to the west. Extending toward its far end I could see Höfn spidering south and outward, capturing the nearby small islands in a web of causeways.

Beyond Höfn spread another wide lagoon, Hornafjörður. Where Skarðsfjörður met Hornafjörður, just south of Höfn, the cradling arms of the dark coastal sandbars opened. There was the narrow gap we had entered, when approaching Höfn from the ocean. Despite the wonderful view, we did not linger for long. Chilled by the breeze, we hustled back into the comfort of the car.

Höfn

Elias's wife had prepared a lavish meal of lamb and vegetables. We lingered with her and her son. "Elias went to Reykjavík with the ambulance and can't join us," she explained. I looked at her quizzically. Was he unwell, or had there been an accident? She caught my expression, "Oh, it's just another of his jobs," she smiled, reassuring. "He's also a customs officer, explosives expert and a part time journalist."

"Why does he have so many jobs?" I asked.

"There are too few people here in Iceland to make jobs full-time. Most people need more than one job to make a living."

That made sense in a country with a population far smaller than that of the town, Brighton, where I grew up.

"Well," said Geoff at length, rubbing his stomach contentedly, "we'd better get going before we take root here!" In clean dry clothes, we offered our thanks and left with Elias's son.

At the water's edge, having crammed our provisions into the kayaks as best we could, we wriggled ourselves comfortably into the cockpits. While we had taken it easy, an ominous grey fog had settled in around the town. None of our earlier landmarks were visible when we drifted from shore.

"Thanks again for everything!" we called, waving one last time before the fog hid all from sight. As we struck out, watching our compasses, I had an idea.

"Shouldn't we be able to find our way just by going with the flow? All this water must be going our way."

The tide ebbing, we had drifted into faster current. Although we could not see our way from here, when I looked at the chart, I recalled the view from above, from the police car. I had seen how the two wide lagoons drained toward the single narrow exit channel. There, the tide would gather speed as the channel pinched narrower, finally jetting out into the North Atlantic. The Admiralty Pilot warns how the tide at the entrance sometimes reaches ten knots. Such current should steer us. How could we go wrong?

"This'll be fun," I whispered to myself, as we cruised along on the swirling water, amid surges and riffles. But then I heard the deep throbbing growl of a powerful engine. There was a boat approaching, and we could not see it. In fact, we could see nothing beyond the small patch of water surrounding us. All around was fog. I looked at Geoff anxiously and he frowned back. What should we do?

ICE LAKE BELOW VATNAJÖKULL.

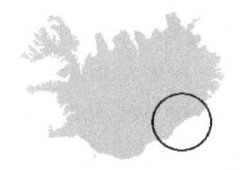

11
Ingólfshöfði

The sound of the thudding engine grew louder. "I think it's coming right here," I said, gripping my paddle, my body tense and ready to sprint, uncertain which way to go. We had best just sit, listen, and hope it would miss us.

"There it is," shouted Geoff. A large wooden fishing boat had suddenly appeared. It swept close by, forged forward into the fog, and disappeared again.

"Phew!" I whistled, "that was scary!" But my relief was short-lived.

"I hear another one!" shouted Geoff. I turned my head to better listen and could make out the rising level of a different engine rhythm against the fading din of the first. My mind flashed back a couple of years to a similar predicament.

It was at night, crossing the English Channel from Dover with my friend Jan. Toward dawn, daylight just beginning to show, we ran into a dense fogbank close to the French port of Calais. It was there we heard the clamor of a ship approaching. It was impossible to tell where the noise came from. It sounded all around us. The thunder of the engines grew louder and louder, while we sat together helpless, tense, and worried, turning our heads, trying to pinpoint where the noise came from. If we were in the way, the ship would hit us. We would not have time to paddle clear.

My stomach felt empty, and my body trembled with the rumble of the engines. Then, it was upon us, the huge bow suddenly in view, the churning bow wave rolling at us. Moments later we bounded through the wake as the side shell plates of the ship streamed past. It had barely missed us.

The ship forged on, swallowed by fog. Behind, the air smelled stale above the swirling, stained, wake. Over the engine's throb I heard a harsh tannoy blaring safety instructions to passengers. A cross-channel ferry had come close to hitting us.

"Here it is," shouted Geoff. I jumped. The second fishing boat was on us. The coarse preview of its painted wooden hull materialized from the mist like a grainy photograph. Looking up to the wheelhouse I saw a pale moon-like face peering down at us from the side window. The face stared, motionless. Suddenly, the boat twisted sideways toward us. It must have hit a sand bank. The power of the current pushed it over until its decks were awash. The engine raced. I sat, stunned. The boat lurched upright again, shedding grey water from its decks and the sound dropped to a slow rhythmical beat. I stared. In moments, the boat faded into a shadowy form and dissolved into the fog. The tide had whisked us away and we could no longer see what was going on.

"Do you think we should turn back?" I called, realizing, even as I spoke, that we would be unable to make ground against such a strong current. And how could we help anyway?

"Was that our fault?" Quite possibly, I considered, but then again, the helmsman could not have seen us until the boat was about to ground.

"I hope they're okay. It sounds as if they are still there." I could hear clanking above the subdued throb of engine. Was that an anchor chain? The sounds grew more distant. We were drifting fast. "Let's get out of here before any more boats come."

Ingólfshöfði

We began cutting diagonally over the waves toward the western side of the channel, assuming the waves had refracted, turning to face the current. Finally, we could make out the shore, and shortly began pitching into bigger swells as we approached the open Atlantic. As we bounded out to sea, in scant visibility, everything seemed surreal.

"Let's keep beyond the surf break," I urged, seeing how rough it was. That plan needed rethinking, for we lost sight of land in the fog. Resorting to compasses, we crept back until we could just see the beach, turning to parallel it. This close to shore, I felt uneasy. From time to time, a critically steep green wall of water burst from the fog, warning us we were too close to the almost hidden beach. This nebulous line was tricky to maintain and increasingly nerve-wracking.

After a couple of hours, we reached a rocky outcrop; a rock trapped by the spread of mainland sand. "Let's stop here," Geoff suggested. Stressed by the conditions, I was ready. We landed and set the tent on a bank of shingle, in preference to the sand, quickly settling in for the night. Although my thoughts dwelled on the fishing boat incident, secure with our replenished food supply I felt ready to tackle the next stage of the journey.

Ahead lay 130 miles of steep sand beaches before the next coastal town, Vík. This was the section of the coast I expected to find the most challenging. "I don't really mind the surf, without the fog," I admitted as I wriggled, trying to bed down the knobby stones under my sleeping mat, "but I don't like big dumping waves, and I read that the beaches along here are steep." Steep beaches cause waves to dump heavily at the shore.

"And I wonder what quicksand looks like. How do you recognize it? What causes it anyway?" I drifted into silence. Geoff was asleep.

We both looked forward to reaching Iceland's shortest river, the Jökulsá, said to be only about four hundred yards long. It was easy to tell we were nearing the river mouth the next day when we began passing chunks of ice, baby icebergs, and we came to land beside where the strong meltwater current gouged a channel through the dark grey sand.

Rounded boulders and stones lay scattered across the sweeping sands as we followed alongside the river upstream. Three hundred yards inland we reached a gravel road, which crossed our path to a narrow suspension bridge over the river. According to my map, this road would continue west before veering inland, northwest, to end thirty miles from here. Current maps did not show the bridges and stretches of road which closed the gap between Höfn and Vík. Since last year, 1976, it had been possible to circle Iceland by road, although judging from the rough state of the unpaved surface here, maybe not in a standard car.

We saw no cars, but there was a rugged bus parked nearby. It had a high road clearance. From the rear wheels back, the chassis rose steeply to allow it to negotiate steep climbs without grounding.

From a short distance beyond the road stretched a serene lake, the Jökulsá lake, Jökulsárlón, filled with icebergs, each a shimmering blue beneath a white top. Blocks of ice the size of houses towered over the water. Everything was quiet, but for the tinkle of falling ice crystals, the gentle dripping of ice water, and the grating cries of terns which hovered, jittering, darting, and dipping to the water. Beyond the lake lay the glacier that spawned the ice, just one of the multitude of rivers of ice flowing inexorably down from the vast Vatnajökull, the largest ice cap in Europe. The chill from the ice around us was palpable. Geoff noticed me shivering.

"If you think you're cold, look at them," Geoff pointed out the ducks. Dwarfed by the chunks of ice around them, they were swimming fast to keep place against the gentle current. "They must be wearing wetsuits," he said, "otherwise their legs would be

Ingólfshöfði

frozen." I visualized ducks wearing neoprene scuba gear, diving to chase fish. How did ducks deal with icy water with uninsulated legs and feet, I wondered?

Passengers from the bus sauntered past, in twos and threes, skirting the lake shore, avoiding us in our strange outfits. From time to time, one would crouch to photograph the ice in the lake, while the others waited.

"I'm still freezing," I complained after a while. "Will you be ready to head back soon?" Our return was brisk, our feet braiding new tracks beside the sand's record of our earlier route.

The coast from here was tricky to read. River mouths did not always appear where I expected to see them. Were they moved by shifting sands, or did I miss something when I consulted the chart? Besides that, the dark sand shoreline seemed endless.

That sand stretched inland too, like a dark grey ocean; its near edge washed by white breakers. Brilliant fans of churning froth raced far up the steep ramp of shore, hovering there before draining back. The retreating foam embroidered a lace of white threads against the dark sand, sucking back into runnels of backwash to greet the next thundering break.

Beyond the coastal plain, like a watercolor in shades of grey, spread a skyline of mountains, tumbling glaciers, and the Vatnajökull ice cap. An insidious icy chill poured silently from the shore, tightening the skin of my face and hands.

Now a headwind sprang up and our bucking kayaks soaked us in spray. Water forced its way up my sleeves until its cold weight swung heavily at my elbows. I paused to drain the water, holding each cuff open in turn, chased Geoff to recover the lost yards and settled back into a steady cruising pace. I had already learned how Geoff hated paddling against the wind, and I could sense his growing frustration now. I kept silent, withdrawn in thought.

Why was I here, I questioned? Would discomfort, challenge and disillusionment make me better appreciate the sparkling times? Must I suffer rain to see rainbows? I recalled my friend Mark Harrison brusquely cutting short someone's complaint, about how hard the paddling seemed, with the curt words, "Sea kayaking is about pain. If you don't like pain, don't go sea kayaking."

Tolerance is key. I had learned long ago how to escape to my inner thoughts, wherever they led, leaving my body to manage discomfort without me. Yet I am constantly surprised how sharp my outward awareness can be, even while I daydream. Who is it that reacts to these ocean breakers, recognizes a speck as a flying bird, notices a change in the cloud, when my thoughts seem to be elsewhere? Can there be two of me?

I tried to ignore how slow our progress seemed, and the cold and wet. Paddling felt undeniably dreary. I watched the shore with little focus. It always looked the same: featureless. But then, I spotted a solitary great skua on the sand, breaking the monotony of the empty beach. As we approached, it fluttered, half-running, half-flying across the sand. It tripped into a heap as if injured.

"Look!" I pointed out. "That bird's injured." I watched it drag its half-open wing along the sand. "Shall we go, and see?"

"Why not," replied Geoff, welcoming the excuse to break the tedium. In unison we sank onto the beach on the back of a dumping wave, where I tumbled face-first into the backwash in my haste to corkscrew, to face the stern to exit. Barely regaining my feet before the next wave thundered in, I snatched my kayak just as the swash caught it and whipped it around, the force almost tearing it from my grasp. I stumbled up the slope after it, chill water pounding behind my knees.

Our kayaks dragged high enough for a quick stop; Geoff ran after the injured bird. As he got close, the skua stretched its wings high, lifted gracefully into the wind and wheeled away.

"Oh well, it gave us an excuse to land anyway."

Ingólfshöfði

We unpacked dried fish, and biscuits, and scaled the sandbank to snack. From there we could see out over the vast, black, desert that lay beyond.

"It seems incredible that all this flat land came from the volcanoes under the ice, doesn't it?" I remarked. According to records, Katla, which sleeps none-too-soundly under Mýrdalsjökull ice cap, erupts every fifty-two years on average and was due to erupt any time now. The eruptions of Katla, in both 1755 and 1918, were so intense that water from the melting ice rushed across the coastal plain at twice the rate of the river Amazon. The debris carried down by each flood extended the coastal plain farther into the Atlantic.

In 1934, the volcano Grimsvötn erupted beneath Vatnajökull, melting four cubic miles (16 km^3) of ice, and producing another incredible torrent of water. Such cataclysmic events, glacial outburst floods, or ice slides, occur often enough here for them to be known internationally by the Icelandic name, *jökulhlaup*. And yet these phenomenal events in Iceland are insignificant compared to others in past geological time.

It was a jökulhlaup after the last Ice Age, 200,000 years ago, that gouged the English Channel, separating Britain from mainland Europe. Others caused at least twenty-five catastrophic floods in western USA, 15,000 to 13,000 years ago. Each of these so-called Missoula floods, in Eastern Washington and Western Oregon, far exceeded the water flow of all the rivers in the world combined.

These floods resulted from ice dams collapsing, each releasing the waters from a massive lake. In one flood, more than five hundred cubic miles of water swept across the landscape in just days. It left a barren terrain so ravaged that, even today, little grows there. Referred to as scablands by early settlers, geologists nowadays know them as the Channeled Scablands.

I shivered, in part from the bite of the wind, but also at my visualization of such a catastrophic event. I hoped there would not be another jökulhlaup here just yet.

"Let's get going." We turned and ran down the steep slopes with long sliding steps, creating avalanches of sand that sped our feet toward the ocean.

It seemed to take forever to reach Ingólfshöfði, *Ingólfr's promontory*. We could see it in the distance, a great chunk of rock; an island now joined by dark sand to the distant mountains. Singing and swearing, and with our bodies aching, we eventually drew near. Noisy clouds of seabirds circled us: puffins and guillemots, gulls, and terns, skuas and razorbills. Seals bobbed and splashed, stretched, and strained to watch.

Seeking the best shelter possible from the wind and swell; we tucked in close beside the rock to land. There, seeing the long reach of the beach break, we thought it prudent to carry our kayaks high up the shore.

"According to the chart, there's a hut here," I pointed out. "Could be convenient, but where is it?"

"On top I expect. Then again, once we get up there, I'm sure we'll see it at the bottom," Geoff grinned. "Let's check it out." We began scaling the steep, seventy-foot-high ramp of black sand piled against the side of the rock. The sand was so steep we kicked steps as if into snow.

The summit of the rock itself was two hundred feet above the sea. There stood the refuge hut, its red-painted corrugated roof bursting from turf walls in the same way that the island seemed to unclothe itself from the encroaching sand. The inside of the hut needed cleaning. Seeing bunks and deciding to sleep there rather than pitching the tent, we began tidying up. It would be a luxury to have so much space.

Our main priority now was to find water. Searching, we located a well. Lined with wooden barrels sunk into the ground, it

Ingólfshöfði

smelled bad, and the water we drew from it looked brown. However, it was free of salt, and both of us had drunk worse. Since we boiled our water for hot drinks and cooking, we thought it would be okay, so we returned to the kayaks to collect what was necessary for our stay.

Gulls and fulmars wheeled around us as we walked, and we became easy targets for great skuas. The body and wings of great skuas are a rich brown color, with just a flash of white on the wings. With a wingspan upwards of four feet, and a bulky body, they appear dark, brawny, and menacing. I was constantly alert for them.

I watched as one skua wheeled around high in the air and swooped straight at me at speed. Its profile seemed small, a dark circle of body with only the leading edge of the wing showing as a line to each side. At the last moment, it shifted its angle of dive slightly. In that instant the bird appeared to triple in size, as the underside of its wings showed, now spread. Its feet dropped and despite my ducking it whacked my head hard as it passed.

Naturally, I kept a close track of it as it climbed and turned, anxious to avoid a repeat, but I was not vigilant enough. Startled by a sudden rasp of wind across feathers, I ducked as another bird whacked my head, from behind.

It was a nesting pair defending its territory; two birds taking turns to attack. If we kept watch on both, they rarely caught us off-guard. However, the rules changed near the border between territories. There, four birds attacked us. The aggressive assaults were so relentless, our only refuge was the hut.

Inside hung a long-handled fowling net, like a gigantic butterfly net on a fifteen-foot-long pole. The skuas had seemed so vindictive, I wanted to use the net to put an end to their attacks. But we were the ones at fault. The skuas were only protecting their nests. This kind of net was for catching puffins.

Iceland by Kayak

Geoff swung the net cautiously in the small room and studied it thoughtfully for a moment. Then, he aimed one end toward the open door and strode out like a lancer. I read his mind. "I'll prepare it if you like," I offered. He looked at me quizzically. "I'll cook, if you get one." He strode off toward the cliffs brandishing the net.

I suppose, like most activities, fowling takes practice. What looks straightforward, when performed by an expert, is seldom easy to master. An experienced fowler casually lifts the net into the air, gives it a twist, and brings it down with a bird trapped inside. That is how fowlers in Iceland and the Faeroe Islands harvest puffins by the thousands. Geoff eventually returned, clutching two herring gulls in one hand and the net in the other. He was scowling and seemed in a dark mood. He threw the net to the ground and shook his head.

"How did it go?" I asked, puzzled. "You got a couple of gulls?"

He swore. "I'm disgusted. Every time I lifted the net, all the birds flew away. So, I thought I'd just climb up a bit of cliff and sneak up to the herring gulls on top." I imagined him climbing one-handed up the crumbling rock, the net in his other. Had he fallen?

"No, I was right at the top when I saw a fulmar sitting on an egg. Once my face came near, it vomited this foul-smelling puke at me." He threw his birds to the ground beside the net in disgust. He looked at me. I looked at him. The bird had scored a direct hit. Seeing the revolting mess in his hair, I burst out laughing.

By the time Geoff had located his soap, he had regained his normal composure. He hurried away, down toward the sea, to wash his hair. Yet despite his best efforts to rid himself of the fulmar oil, it left a lingering pungent stench that persisted for days.

If I laughed hard, the last laugh was on me. As I began to pluck the first bird, my damp fingers became clotted with clinging

Ingólfshöfði

feathers. Pale clouds of down circled around me, gently floating until finally alighting on the grass, and on me, like paper ash. I felt as if I was inside a shaken snow globe. I stopped and thought for a moment. It would be easier to skin the birds.

So, I cut off the head and began to run my fingers between the still-warm carcass and the skin, preparing to ease the skin off in one piece. Suddenly, the body in my hands let out a loud vibrating cry. I dropped it in horror and stared in disbelief. Surely, it was dead, this headless carcass.

Gingerly I picked it up again, to resume where I had left off, and jumped again, startled by another loud cry. I was so focused; I was unaware of another sound. I turned to see Geoff collapsed on the grass, convulsed in laughter. Had he been playing a trick on me?

No, I realized. Beheading the bird, I had left the neck and syrinx attached to the body. The pressure of my hands had forced air from the lungs, making the corpse squawk.

Later we reviewed what was, for us, a new dish: curried herring gull with rice and chapattis. "Not bad," considered Geoff savoring the flavor. "No, that's not bad at all!"

It was at this mealtime that we began having difficulties with our tiny petrol stove. It would not burn properly, yielding a weak yellow flame, with a sickly-sweet smell, instead of a roaring blue flame. "I'll look at it," Geoff said, taking a pair of pliers in the absence of the correct spanner. The brass nut resisted. The steel teeth of the pliers simply burred over the edges of the nut, sprinkling golden filings.

"Be careful with it," I implored, but Geoff had the bit between his teeth. He had other tools from the hut. Taking a hammer and a screwdriver, he pounded at the nut until it succumbed and began to unscrew.

"Just look at this!" he groaned in revulsion, pulling out the wick. If it had once been white, it was no longer. The tarry black worm of wick slithered from the stove onto the grass in front of us. "Look, it's all clogged up in the pipes too." Geoff began to clear them as best he could and rinsed the wick in gasoline. One piece at a time, he reassembled the stove. Now, when lit, the stove burned with vigor. Geoff had remedied the problem, but what had caused it?

By the next day we knew the fix had only been temporary. Geoff, in disgust, pounded the stove open again, dragged out the tarry wick and soaked it again in gasoline to clean it. I tried to pry the tar from inside the jet and the pipes. "It's not going to be easy if we can't use the stove along this next section," I frowned. "We'll have to cook on fires, and there'll be no shelter."

As we reassembled the stove, we heard a small plane flying close and rushed from the cabin to see. It circled low over the rock, dropped, passed the island again below the level of the clifftop, and headed away west following the coast. We were surprised, jolted from our solitude. Geoff finished reassembling the stove and it lit with a reassuring roar. Relieved, we brewed mugs of tea and cooked ourselves a meal. Everything was rosy again.

***

I stood alone on a high corner of the island, looking out across the dark sea of sand toward the distant mountains. Turning, I saw the vast corrugated surface of the Atlantic Ocean. I was on the bridge of my own ship, high above the water, dragging a sheet of sand behind me from a far shore. I felt a snug security here, the air vibrant with the frenetic activity of thousands of sea birds. Sheep scrambling over the metal roof had awoken us this morning. Below me now, seals cavorted in the bottle-glass sea.

Ingólfshöfði is known after the Norwegian, Ingólfr Arnarson. He landed here around AD 874, before becoming one of Iceland's first settlers. I could imagine him standing here all those years ago, this rugged oasis his temporary refuge. His foster brother,

Ingólfshöfði

Hjörleifr, landing farther west, did not survive for long. His ten Irish slaves killed him, and his men. They took his boat, with the women, to a nearby island.

Finding Hjörleifur's body the next spring, and seeing his boat missing, Ingólfr correctly guessed where the slaves had gone, and confronted them. According to the sagas, those who chose not to face the sword jumped to their death from the cliffs. The islands have ever since been known as the Vestmannaeyjar, *West Man Islands,* the west men being the Irish.

Ingólfr eventually settled in the west of Iceland, at a place now part of Reykjavík. We would follow in the same direction. Ingólfshöfði is the last haven before the next coastal village, Vík, eighty-five miles west. Between here and Vík lay sand. Black sand, with no natural shelter. Apart from occasional shipwreck shelters, it is unshaped by man and seldom visited. With my feet grounded on lichen-crusted rock, I savored the view with apprehension.

BLACK SANDS, FROM INGÓLFSHÖFÐI.

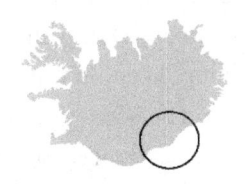

12

Into the Sands

Huge rafts of guillemots parted as we approached, skittering aside, or ahead, to reform in dense groups again. The whirring of purposeful wing beats came from others crisscrossing in flight all around. Puffins too cut paths across the sky or floated high on the water. Dark shapes winged beneath us as birds escaped underwater rather than lifting into the air. In contrast, calm, dark-eyed fulmars regarded us as they circled effortlessly past, almost brushing the water with their wingtips as they swept close by. A thousand murmuring voices broadcast from the cliff.

We rounded the last corner of rock and saw ahead the dark sands extending into the hazy distance. As we drew away from Ingólfshöfði, where the air and water bustled with seabirds, we saw fewer birds, until eventually we were alone. The grey sky, grey sea, and grey sand, all blended into a single blur in the distance.

We began to paddle by the clock, for two hours at a time before rafting our kayaks for a short stop, with a bite to eat. In that way we gauged our progress along the featureless shore, paddling toward a time goal rather than a distance or destination. As I paddled, Donovan's song, *Sunshine Superman*, cycled endlessly in my head, in time to my paddle strokes.

The scenery did not change. Our charts continued to promise river entrances where we saw no sign of them. Had the sand shifted, blocking them, or had the rivers dried up? The only definitive landmark we passed was a shipwreck shelter. Now, after six-and-a-half hours of paddling, through light rain, Geoff spotted a square object speeding along the beach.

"Can you see it?" he asked. "What do you reckon it is, a Jeep perhaps?"

I could not tell. It looked as if it moved very quickly. We watched intently as we drew steadily closer, in the end deciding it could be a shipwreck shelter.

"I think it's an illusion. The sand is so featureless, and the clouds and surf move so much, it looks as if everything moves. Nothing stands still." Without any reference against which to judge distance, or size, in such a blank landscape, we were still uncertain what we saw even when we drew quite close. Nevertheless, it was encouraging for if this were the second hut, we should reach another in about an hour. Then, we could pinpoint our position on the map at a good place to stop.

Paddling was turning out to be less tedious than I had feared, despite the weather turning to heavier rain. The rhythmic *Sunshine Superman* continued to skip along in my head, and occasionally I burst out into song with the few words I knew. Geoff joined in, and we sang a raucous duet into the vastness of the grey day.

At half past midnight we spotted the hut we were looking for. It seemed to race along the sand like the earlier one. The surf looked powerful. Sitting beyond the break we could make out the effect of what we assumed was a submerged sandbank. It caused heavily breaking waves which re-formed into smaller swells where the water deepened again beyond the bar.

Into the Sands

We skirted around the end of the bar, breakers rolling in from our left, and then cut inshore of it, sneakily avoiding the full power of the surf. Ploughing up onto the sand, we scrambled from the kayaks, relieved to be ashore at last. However, the hut was on the far side of the river that had created the bar.

We stood and watched the river churn past. It erupted everywhere into great boils of dark, grey, sandy water as it muscled powerfully down the beach. Gingerly assessing the depth, we discovered the current had scoured the bottom into an egg-carton pattern, with deep hollows, and ridges. The strong current would sweep us away if we tried to wade. Any attempt to kayak across the entrance, or around the bar, would deliver us into the worst of the breakers, those we had sneakily avoided.

Instead, we carried our kayaks far upstream to launch. Snatched by the swirling current, the kayaks swept rapidly toward the ocean as we sprinted for the security of the far bank.

Ashore again, we wondered where to leave the kayaks. Not only did the dark sand stretch as far as we could see along the coast in both directions, but it also stretched just as far inland too. Great bleached logs, and green glass fishing floats, gleamed in the rain. Wooden fish boxes and other debris lay scattered as far as the horizon. How far up this beach would the sea reach at high tide? What if the river rose? Should we carry the kayaks all the way to the hut?

We carried the first one a long way, but since the hut was farther away than it had originally appeared, we stopped short and retrieved the second. "What if the wind rolls them, once we've taken our stuff out?" It was a risk. Geoff pounded a driftwood stake deep into the sand, tied the kayaks to it, and lashed the kayaks together for good measure. The stake was a token gesture, for nothing would hold for long in sand. Wearily, we trudged through the rain toward the shelter.

Iceland by Kayak

The hut was a neat square structure, painted in bold red, held above the ground on wooden supports to prevent sand from drifting against the sides. Far from collecting sand there, the wind had instead scoured a hollow underneath, where a pile of plastic debris had gathered.

Climbing the open wooden staircase, we swung wide the outer and inner doors and stepped into the stillness of a single dark room. As our eyes grew accustomed to the dimness, we could make out a couple of tables, benches, folding beds, survival rations, first aid kit and bible. These were for use by shipwrecked mariners.

I imagined a fishing vessel riding out a storm in the long dark night of winter, straying toward a shore so low and flat that even radar cannot warn of its lethal proximity. Dozens of such vessels lie buried in the ever-encroaching sands. I wondered how many unfortunate crews survived the trauma of shipwreck only to perish on shore without shelter.

It was the frequency of such disasters, with often whole ships' crews huddled together on the sand, dead from hypothermia, which eventually prompted shipping companies to act. They funded the building and provisioning of refuge huts on remote parts of the coast. These have since paid for themselves, repeatedly, by saving lives.

There was a map posted on the wall. Geoff struck a match, lifted the glass on a paraffin lamp, and lit the wick. In the warm glow, we could now see a layer of grey dust coating everything. We also saw how the map showed the position of the hut on the coast, and a route to the nearest farm, eight miles inland.

As I stood there in the calm, every bit of my remaining energy seemed to drain away. It took all my willpower to step outside, shut the doors to the cozy shelter, and trudge back to fetch the necessary gear for the night. Afterwards came the reward: dry

Into the Sands

clothes, and a meal in the comfort of the hut, knowing we would sleep there.

The following morning, we flung open the double doors to reveal a dismal scene of driving rain. The surf thundered much louder outside than in, and the low scudding cloud, together with the dark sand, seemed to swallow the daylight. Rainwater cascaded from the roof.

"Well, at least we can fill our water bottles," Geoff suggested. We lined up our bottles and spare pans under the curtain of water. Retreating inside, we shut the doors against the noise while we cooked breakfast. Nourished, and with our water bottles full, there was nothing but reluctance to keep us from leaving.

"Ooh, this feels gorgeous!" I shouted as I pulled wet, sandy, fiber-pile clothing onto my till-then warm dry body. In truth, I detested mornings like this when my clothes were cold and wet. And this was just the start. If it continued to rain like this, I would be wet to the skin for the rest of the day, and cold too.

My most acute discomfort now over, I stood and taunted Geoff while he suffered the same spasms of self-torture. Both ready, we gathered our bags of gear, shut the doors behind us and meandered through the driving rain to our kayaks.

Waterlogged sand packed the cockpits. "Do you think the tide came up this high?" I questioned.

"I don't know. There's an awful lot of sand in mine."

When I tried to tip my kayak, to drain the water, I could not budge the weight of so much wet sand. I pulled the hood of my anorak over my head against the rain, knelt beside my kayak and scooped out the sand through the water, handful after handful. My fingers hurt with the sharp grit under my fingernails, my hands burned cold, wet, and gritty. It was raining hard, and windy, yet we had no choice. No matter how uncomfortable we got, we could not move the kayaks without first emptying the cockpits.

Iceland by Kayak

The river had swollen in the night and was now a raging torrent. Line upon line of standing waves appeared to run upstream without making headway. The ocean was rough. I could not see beyond the first few massive lines of surf.

"What do you think?" I asked.

Geoff shrugged, "It looks pretty substantial."

"Do you think we could use the river to help us get out?" The current rushing out through the surf would surely help us break through the waves. "It doesn't appear to be damping the surf down though," I added.

I scanned the surf, looking for the telltale path where the current should disturb the pattern of breaks. Such irregularity should reveal the easiest route to open water. I expected to see it, especially with a current this powerful, but I could not make it out. "The surf looks the same all the way along the beach," I concluded. "It's big, and chaotic." I grew unsure of myself now, yet a current this strong must help carry us out to sea. I made up my mind. "Let's start in the river anyway. It's got to be the easiest way."

We launched from a steep-cut wall of sand into the river and let the rushing current sweep us down, straightening up just in time to pound head-on into the surf. For a moment all went well, before I began to struggle. Each successive wave reared taller; the great grey walls thundering at me with startling power.

We were still far from reaching the break line, and open ocean, but the farther we got the more violent the breakers. I piled into a massive roller, so heavy I could not push through. Forced backward with my bow pointing skyward I twisted into a sideways brace, scared I might collide with Geoff until I saw my way clear. By the time I freed myself from that wave, I was back near the beach again, having lost my hard-earned ground.

I began forging out again, but the steep crashing waves broke too closely behind one another. There was not enough time to

Into the Sands

recover from one wave before the next avalanche thundered down, too little time to push forward between breakers. My kayak exploded through one crest, lunged into the air, and crashed down into the base of the next gnarly wave, its steepening wall about to collapse. Trapped beneath the falling water I clung onto my paddle as the roller bounced me shoreward again.

Geoff, in view for just a moment, was farther out than me, clearly struggling to hold his position. What would we do if one of us made it out beyond the break, and the other could not? There would be no way to communicate. Before, when standing on the beach, we had been unable to see the horizon beyond the breakers. Would the shore be in sight, viewed from the ocean?

By now I seriously doubted I could penetrate the surf today. What would the sea be like farther out anyway? It would be crazy. I saw Geoff start up a wave, just as the top began to collapse. Climbing into the breaking crest, he disappeared. Long moments later, the point of his yellow deck launched skyward from the foam as he pitched backward, end over end. Then, the waves hid him from view.

Long slow minutes seemed to pass before I next spotted him, clinging to the end of his upturned kayak, swimming for shore. For a scared moment I wondered if the current would carry him away. Looking around, I saw with relief how far the wind had already swept us along the beach, well beyond the current's reach. I maneuvered closer.

"Grab that bag for me, will you?" he called. There was a large bag floating nearby. I snatched it up onto my lap and attempted to stall nearby until Geoff reached shore.

"Sod this for a lark," he said as we stood shivering, watching the gnashing chaos of waves. Rain and spray stung our faces and rattled against our anoraks. "Let's wait till it calms down a bit."

Trudging back toward the hut, Geoff explained, "I made the mistake of stuffing my camera bag between my legs. When the wave took me over backward, I hung on underneath until all the turbulence stopped. The moment I rolled up, the next wave bowled me over again. Hanging under there, I was running out of air, and the camera bag had moved. I was worried I would be unable to get out, that it would trap me, so I bailed." With such a small cockpit, I was glad he got free.

The windowless hut seemed even more inviting now. Back in the soft lamplight, with a cup of hot chocolate between my stinging palms, and the roar of the surf now muffled, I sat in utter contentment.

"Lucky you were clear of the river." I voiced my fear, "I was afraid the current would carry you out to sea. Thankfully, the surf washed you back in, no damage done."

Unlike me, Geoff seemed happy when swimming. This reminded me how our mutual friend Jan had mentioned, in passing, a scary incident Geoff had on an earlier expedition. I never heard the full story, so I asked, "Geoff, didn't you have a long swim on your solo trip around Britain? Jan told me. Four miles, or something, at least I think that's what she said?"

Geoff laughed. "Yeah, it was a little bit tricky. I was crossing Solway Firth from Scotland to England, about sixteen miles. It wasn't the worst day, not the weather anyway. There was a heavy swell, not breaking all the time but enough to make the kayak plunge. Each time the water washed over the boat, a little leaked in through the spray deck."

"An Angmagssalik, wasn't it?" I asked. The kayak took its name from the town on the east coast of Greenland where the design originated.

"Yeah, my woodwork teacher at school thought up the idea of building kayaks from plywood, stitching panels together and sealing the seams with fiberglass. *Kayel* construction. Sailing dinghies, like the Mirror class, use that method now." he

Into the Sands

explained. "That's *Kayel* from his initials, *K.L.* for *Ken Littledyke*. It's a neat way to build boats.

"I remember when Oliver Cock came in one day." Oliver was the National Coach, director of coaching for the British Canoe Union. "Oliver had this sealskin-covered kayak from Greenland. He wanted my teacher to make a plywood replica. The Angmagssalik is a direct copy of the outside shape and size.

"The panel shapes were easy to copy, although when you sit on the water in a skin kayak, the skin sags in, so you get all these concave surfaces, which add stability. The plywood version is tippier than the original because the plywood is stiffer.

"A mate of mine built one, so I wanted to do the same. Then, I decided to make a trip in it, though I had no idea where to go. One day I thought, why not paddle around Britain? So, I did."

I queried, "It's not a very big kayak for that is it?" I had paddled one the previous year, 1976. Getting into it was like pulling on a pair of tight jeans. The circular cockpit angled up from the flat rear deck, which allowed me to sit on the back deck and wriggle forward, feeding my legs into the tight space until I dropped into the seat. Just nineteen inches wide, with an extremely low profile, the Angmagssalik was nineteen feet long including the slender point at each end. On the water it balanced like a floating needle. Even on flat water, the back deck was awash. "That thing looked ready to skewer fish for a barbecue. It had about enough room for a sandwich in each end, nothing more."

"Oh, it's not as bad as that," Geoff chuckled. "Although she's no cargo ship! I carried my tent, sleeping bag, clothes, and a stove, but I didn't camp every night. Sometimes the Coast Guard called ahead, and somebody met me at the end of the day to offer me a place to sleep. They were extremely kind."

"So, what happened in Solway Firth?"

"My kayak started filling up with water until, by the time I was about halfway across, the cockpit was half-full, and the kayak

Iceland by Kayak

was wallowing. It was getting so sluggish; I was scared the hatches were leaking too. It was a misty sort of day, paddling by compass, out of sight of land. I was seriously worried I might sink. I tried to bail the water out using my hat, but with the cockpit so low to the water, I was shipping water faster than I could bail it.

"When I saw chimneys ahead through the mist, I grew more confident. Suddenly, the kayak wallowed side-on to a wave and all the water inside shifted. I braced, my paddle snapped and over I went. With the kayak so deep in the water, my body would not sink underneath, so I could not get around to my good side to roll. Of course, I only had the two broken pieces of paddle anyway. I bailed out.

"I got myself sorted, stowed away part of the paddle, climbed back in from my good side, and rolled up with the other piece of paddle. It was tricky to balance a swamped kayak with half a paddle in that sea. The bow was so full of water, the kayak wanted to turn on every wave, and it was challenging to steer with only one paddle blade.

"After a while I spotted a buoy and headed for that, but the waves made the kayak lurch away in one direction, and then another. I was trying to make progress, steer, and stay upright, with half a paddle. In the end I decided it would be easier to swim and tow the kayak. I tied the end of the painter around my waist, to free my hands. It took me the best part of an hour to reach the buoy.

"I tried to climb up onto it. That was a game," he rolled his eyes. "It was hideous, wrestling with a big slimy tin can, which swung around, and leaned toward me every time I grappled with it. It was a beast!"

"How on earth did you climb up?" I asked.

"Well, there was a current, so I swam around the buoy to loop the painter around its mooring chain, with me on one end and the kayak on the other. Then, I tried to flick the line up the side of the buoy, to catch the metal handle on the top; the handle they use to lift the buoy onto a boat when they take it away for painting. Anyway, the rope kept slipping off, so I would start all over again, swimming around and flicking the line, until eventually it caught on the handle. When I tried to climb up, using the rope, the buoy turned. It leaned over me whenever I put my weight on it. It was a pig of a job to get up. But even once I was on top, hanging onto the handle, I could never sit upright. The buoy always leaned under my weight, and it leaned me backward.

"When ships passed, I tried to catch their attention. No luck. It got dark. In the morning, after I had spent the whole night clinging to the top of the buoy, I realized the line holding my kayak had snapped and my kayak had gone. No one was going to find me. I waited till the tide stopped going out and swam to shore. I was half-dead by the time I got there. Without my shorty wet suit, I would have frozen. Then, someone took pity on me and offered me a hot bath."

"Whatever happened to your kayak? Did you ever find out?"

"Yes, it washed up in the end. The kayak was a wreck, but I got all my gear back. I thought, well, that is the end of it. But after I had been home for a while, I changed my mind. My mate Ian had a kayak the same as mine. I beefed his boat up a bit and finished the trip in that."

It was an inspiring story. No wonder Jan was impressed when she heard it. I refilled the stove for another brew and thought of all the ways in which this trip differed from his earlier one. Firstly, there were two of us to deal with tricky situations, and now, seven years after his trip, we had kayaks designed for expeditions, rather than for hunting seals. With deck-mounted bilge pumps, there was no need to take off a spray skirt to bail. We also carried spare paddles.

Geoff's exploit reminded me why I invited him to join me in the first place. He was amiable, self-reliant, and resourceful. In a tricky situation, likely to make a good decision he would be open to alternatives and would keep trying until something worked. He was a survivor.

Settled into the hut, I could still hear the muffled roar of the surf even through the sturdy walls. Although we had no intention to leave just yet, I grew anxious. What would the surf be like tomorrow? With little food left, what if we got stuck here?

CAMPING ON THE SANDS.

13

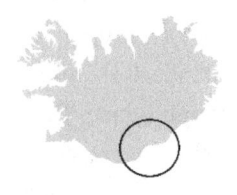

The Long Wade

Next morning, I awoke to a burst of wind and saw Geoff hurrying out through the door. After a while, he returned to say he had suddenly worried about our kayaks. "I wondered if we had carried them far enough," he explained, as he leaned against the inner door, pushing it shut. "They are still there, full of wet sand, and the surf's still huge."

We cooked porridge and then stuck our heads outside again to check the weather. The wind, which had increased, now blew from the south-west. Our guessed weather scenario was a low-pressure system passing to the north, but we had no way to get a forecast. By the time we were ready to change for paddling, the wind was even stronger.

"I think we should stay put until it blows over," I suggested. "It'll be even worse out there than yesterday." I felt relieved when Geoff agreed. Yesterday's fiasco had shown we were no match for these conditions. We busied ourselves cleaning the hut.

I studied the escape routes shown on the wall map. We had little food left, and I worried we might run out before we reached

Vík at the end of these sands. We had already eaten all the dried food I brought from England; all that the rodents in the Faeroe Islands left untouched. Little remained of the bulkier food we bought since. Rather than spend the entire day here, I decided to follow the route marked on the map to see if I could find a store.

"Do you want to come?" I asked Geoff.

"Nah, I think I'll hang out here," he said at first, but with persuasion he reluctantly agreed to come.

Leaving at around one-thirty in the afternoon, we followed a line of marker poles that led from the hut. The landscape was bleak. The strong wind coming off the sea carried drifts of dry sand past us, filling our trainers with sharp granules. Here and there, bleached driftwood, smooth from years of sandblasting, held gleaming limbs to the sky.

Approaching a barrier of dunes, we saw how coarse tufts of grass clung half-buried along the edge where nothing else could grow. The huge hummocks beyond must have begun life in this way, the gradual stabilization, and trapping of the sand, allowing these green hairy mammoths to grow twenty feet high.

We continued to follow the markers between the dunes until our way opened onto the shore of an enormous lake. It was so wide we could see no end to it. The low cloud reduced the vista to a thin horizontal slit between the cloud and its mirror. Across the lake, and into the distance, stretched a single line of forty-gallon oil drums, each painted red and yellow. "Wow, this looks weird!" I exclaimed.

"Well," said Geoff, "either the water must be pretty shallow, or they've stacked up the drums, one on top of another!" I laughed, visualizing that absurdity.

I pulled off my shoes and socks and rolled my jeans up to my knees. Wading barefoot into a full half-inch of pellucid water, I turned with a tongue-in-cheek, "It's deep here." Geoff rolled up his trouser legs and followed. We splashed through the chilly

The Long Wade

water for an hour or so, chatting. Nowhere was the water deeper than a couple of inches. We stopped to look around us.

"It's just as if we're standing on top of a huge lake." Water stretched into the distance in all directions. We must have waded two miles. All I could see was this row of well-spaced painted drums dividing the sheet of water. The effect was bizarre. "You could probably get an Arts Council grant to create something like this," said Geoff, "not that anyone would come all the way out here to see it."

The water deepened incrementally. After hours of wading and with the water at our knees, we had come so far there would be no point in turning back. We paused a moment to discuss the shadowy smudge of what I took to be a distant bank or shore. "It's got to be the other side," I said hopefully. "My legs are frozen." I longed to stand on something dry, to warm my feet. "I had no idea it would be like this."

We pushed on, swinging our legs gently forward through the water, trying to keep our jeans as dry as possible. When we did reach the bank, we were disappointed. Far from being dry land, it was a waterlogged mass of moss and grass, dissected by myriad wandering waterways, sandy-bottomed. A solitary sheep stared at us from a spongy hummock. We squelched past, warily avoiding the hidden hollows into which we occasionally stumbled up to our thighs.

Eventually, we reached a tiny house and knocked, intending to ask where the nearest shop was. The elderly couple answering the door looked horrified, if not terrified. Remorseful, I realized how disreputable we must look. Our jeans were mud-smeared and soaking wet, and we both had ragged hair. This was a secluded place, at the end of a long gravel road that led from who knows where, and they could see we had no transport.

"Hello, do you speak English?" I began. Perhaps if they had, or had we spoken Icelandic, they would have asked the obvious questions: "Who are you? Why are you here? How did you get here? Where is your car? What do you want?" As it was, we could no sooner understand their questions than they ours. The old lady ushered us inside and shut the door behind us. We stood awkwardly in the narrow hallway, wondering what to do next.

Suddenly, the lady's face brightened. She picked up the telephone and spoke for a moment or two. Handing the phone to Geoff, she rolled her hand in encouragement. Geoff held it tentatively to his ear and asked cautiously, "Hello? Do you speak English?" He broke into a smile, nodding his head at the old lady. He cupped the mouthpiece of the phone to confirm to me, "Someone who speaks English."

He began explaining on the phone who we were and what we wanted. He passed the phone back to the lady who, as she listened and talked, began to relax. She returned the phone to Geoff to listen, and he explained the plan to me afterward.

We were to wait there until the postman arrived. He would take us to the main road, where a local woman would pick us up.

Meanwhile, the old lady, with great courtesy, ushered us into the kitchen, where she prepared coffee and cake for us. We had only a short wait before we were away, rattling down the track in the postal van, through an ancient lava field. The lava was so old that the contorted shapes had become draped in long grey-green fleeces, the lichen dangling in shaggy folds. We pulled up abruptly, in a cloud of dust, at a junction.

"Wait here," the postman said, "someone will come." We thanked him and slammed the doors shut. He sped away, trailing a plume that lifted, and spread, marking his progress far beyond where we lost sight of him.

"Well, that's that!" I said, looking around. The graded road stretched in both directions, potholed and rocky, with no sign of any car. "What if nobody comes?" I looked down at the damp denim clinging to my legs and thought about the long hike we still

The Long Wade

had ahead of us, wading back to the coast. Something seemed wrong here, provoking a sinister feeling of déjà-vu.

Our situation reminded me of Alfred Hitchcock's American spy movie, *North by Northwest*, when Carey Grant, as the fugitive Thornhill, gets off a bus at a remote crossroads. There he expects to meet an agent, Kaplan, who unbeknown to him does not exist. Once the bus leaves, he is alone, just like us, with no sign of anyone around. That is except, in his case, for a distant crop-duster plane, which eventually attacks him.

"There's a car," said Geoff smartly. I looked up. The car sped toward us trailing a cloud of dust as dense as from any crop-duster. It weaved from one side of the road to the other to avoid potholes and rocks.

Finally, the battered and dusty Sunbeam pulled up beside us, where the vivacious lady driver urged, "Come quickly! Get in, we don't have much time!"

As she steered adroitly, selecting the least punishing route between holes and rocks, smaller stones kicked up by the wheels smashed violently against the chassis. With each impact, a cloud of dust exploded from the floor around our feet. Is this really the main road, the ring road around Iceland, I wondered?

Despite our haste, we were only just in time. At the store, the shopkeeper was already counting her day's takings, cashing out. She asked us to be quick, so we hurried, grabbing essentials as fast as we could. We were soon back on the road again.

Expecting to be dropped off where we had been picked up, we were surprised when the woman, driving straight past the turn-off, said matter of fact, "Of course, you must come with me for something to eat. Later my son will take you back in his Land Rover."

She collected her son, Bjarni, from the school where he was employed with her husband and took us all to their farmhouse. There she fed us, and before we left, presented us each with an

Iceland by Kayak

Icelandic sweater she had just finished. She had knitted them in a traditional circular yoke pattern from soft Icelandic wool. I slipped mine on, lightweight and luxuriously cozy, and admired its natural shades of grey and white. I felt honored.

Later, we climbed into the open back of Bjarni's Land Rover and clung on while he drove us back toward the coast, crossing the lagoon by an unmarked route. He stopped to check the depth wherever he was uncertain. Wading ahead he probed the water with a long stick. It was like traveling by boat, except taking soundings in case it became too deep, rather than too shallow.

When we reached the hut, and we had shown off our kayaks, Geoff, in gratitude, presented the lovely lady with a small bottle of brandy. He had carried it to celebrate the completion of our journey, should we succeed. She seemed thrilled.

The following day, the wind had shifted to the southeast. We surveyed the shore to find the best launch spot. There, we waited for a lull, sprinting from shore to avoid a pounding from the waves of the following set. I was lucky, but Geoff less so. I scanned behind me from the top of each swell until he burst into sight through a wave. Soon he was alongside. We were both soaked, so my remark, "You look a little damp," was more to question why his hair stood on end sideways in a clump.

"Same stupid thing as last time," he replied, "only this time I rolled up, no problem."

Refreshed, cruising over breathtaking swells and optimistic, we resigned ourselves to a less perfect outlook when the wind swung around against us. Geoff reiterated his dislike of headwinds, so when in the early afternoon we spotted a square shape on shore, which we took to be a refuge hut, we chose to land. Although there was a river outlet nearby, which I thought confirmed our location, there was no sign of a hut on the chart.

The surf smaller here, we landed easily. After carrying our kayaks well clear, we looked around to find the hut had vanished.

The Long Wade

"It must be hidden by those sand dunes," we concluded. Searching, we realized there was no hut. In its place stood a massive bone, part of the skull of a large whale. We burst out laughing and walked around and around it. With nothing to gauge size against, we had assumed it had been a hut because of its regular shape.

Now we scrambled onto it and ran our hands across its warm irregular smoothness, feeling its slight oiliness. The square upright part stood taller than my waist. Stretching out from that were two long curving slabs of bone. At a guess, the bone measured more than ten feet long. I was fascinated.

Resigned to the reality of no hut, we fetched food and sat side by side on the "settee," our legs stretched out along the horizontal bone. Our backs, sheltered from the wind, leaned comfortably against the naturally insulating upright.

Having eaten and rested we pinned the tent in place with logs. There were plenty to choose from, scattered across the sand. Given such abundance of wood, we scooped a hollow in the sand and stacked a pyramid of dry sticks. In no time, Geoff, dodging to avoid the wind-fanned flames, began cooking a spicy curry with countless pancakes. Curry and pancakes had become our easy-option meal.

After our nights in refuge huts, pitching the tent brought us right back into the elements again. The breeze flapped the tent fabric, flicking sand everywhere. When I stood out of range to finger the grit from the corners of my eyes, I felt at ease and comfortably warm. The sun-heated black sand pampered my bare feet.

Arriving here by kayak made me realize how vast and empty the surroundings were. There was a faded distant view of mountains, ice caps catching the sunlight. Dark streaks, here and there, showed where bands of rain fell. The wilderness of coastal plain spread all the way from here to those mountains.

I drifted around, wiggling my bare feet down into the warm sand, picking up cartons and bottles to see where they had come from. From the languages on the labels, these were from Europe. Geoff assured me that the glass fishing floats scattered everywhere came from Portugal. Driftwood logs lay strewn across the landscape, far into the distance. I lifted a glass float, green against the light, and set it gently back down. It was not something I could carry home from here. A compulsive beachcomber, I live in the hope of finding something small of interest or value. Mostly I find something useful, such as a wooden fish box to sit on, or driftwood for a fire.

Next morning, I drifted into wakefulness, aware of early morning rain. No, not rain, I realized; it was a warm morning. Condensation in the tent was flicking down on me. The surf crackled, rumbled, and hissed, creating a background wall of sound. I stood outside to assess the wind, and the sea state, and judged it to be about the same as yesterday. It was only six o'clock in the morning, so I was in no hurry to move. I ducked back into the tent and fell asleep again.

"Let's have a lazy day!" suggested Geoff later. "We can get an early night tonight and leave first thing tomorrow. It'll be fun to hang out here." I agreed. So, at our little fireplace, I dug into the ash where I buried the fire last night. Scooping out the hot charcoal embers that still smoldered beneath, I soon re-kindled the fire.

In the sun, dancing heatwaves blew across the dark sand making the view of the sea tremble. The flapping tent hissed as drifts of sand rattled against the fabric. Trying to chill out I could not sit and relax. Close to the ground wherever I sat, inside or outside, sand blew into my eyes.

To escape that airborne irritant I wandered inland, finding ever more glass fishing floats. Some were the size of a grapefruit, others as big as a football. As I strayed, I realized how uniform the

The Long Wade

landscape appeared, even far from the ocean. It was a single endless dark sand beach. All around lay more driftwood, floats, and wooden fish boxes. I could imagine the surf repositioning everything daily. In truth it might have lost interest in this flotsam and jetsam decades ago.

Having roamed far from camp, I turned from the view of distant mountains and meandered back toward the tiny orange target of the tent. It was time for another mug of cocoa.

When the next day continued windy, we procrastinated. There seemed no good reason to push on despite. Deciding it would be better to enjoy our time here, we changed our minds when the wind eased in the afternoon. We set off happily until the wind picked up strongly in our face again. It made us strain the last few miles to reach a refuge hut which, close to shore, was clearly not a whale bone.

Our plan to cook in the hut stalled when the stove malfunctioned again. When Geoff began to dismantle the stove, I escaped to fetch fresh water from a little delta I had seen draining into the surf. Our reward for Geoff's success was a pot of hot soup. We slept soundly in the shelter and by next morning, to our delight, the wet gear left outside had dried.

On the ocean, feeling not a breath of wind, we began to spot seals again. It was a welcome change, for along the sandy coast since Ingólfshöfði we had seen little sign of any seal, seabird, or even land bird. The scenery too was changing; the mountains drawing closer. We began heading toward cliffs instead of that vague, distant smudge where the low sand beach appeared to fade at the horizon. We were approaching the small town, Vík. A vík is a bay. To distinguish it from other bays, people know this settlement as Vík í Mýrdal, *bay of mire dale,* or *moorland valley bay.* I felt a growing excitement, like the anticipation when approaching shore after an open crossing.

AUTHOR RELAXING ON WHALE BONE.

GEOFF WITH WHALE BONE.

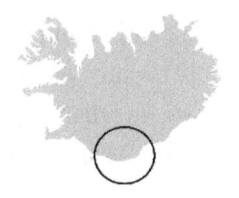

14
Vík

As we neared Vík, a series of weirdly shaped rocks became clear in the distance. "Can you see the one that looks like a bust of Neptune?" Geoff asked. I could, and the closer we drew the more realistic the features appeared.

The town hid itself mostly from view beyond the steep black sand beach, across which the surf thundered. There did not appear to be an easy place to land.

"Right over there, at the far end, by the rocks?" I suggested. "We might get a bit more shelter there." We pushed on to where huge and vegetated rocks reached down to meet the end of the steep beach. The shelter was relative, for to make it ashore required careful timing, riding on the backs of dumping breakers. Relieved to have landed without mishap, we stood there grinning at each other.

The sky around us pulsed with energy, a relentless frenzy of wingbeats as puffins, guillemots, and razorbills shot by. Terns and gulls shrieked all around. This was breathtaking, even more so in contrast to the sand-dominated landscape of the last few days. Then, the only sounds had been the roar of the wind and surf, and the hiss of wind-blown sand. For days we had seen nothing nearby of elevation, just low-lying scenery with occasional glimpses of far distant mountains and glaciers. We seldom saw any birds.

The huge green cliff above us, steep, with the top hundred feet or so overhanging, was home to myriad seabirds. "This is brilliant," I burst out. "I've never seen so many puffins!" And it was not just the puffins that made me joyful. Having seen scarcely any plant life for the last days, it was uplifting to see lush green vegetation smothering the rock. Above the foreshore a barrier of knee-deep, blue-green, marram grass contrasted the black sand.

Leaving our wet gear spread in the sun, we gathered the letters we had written and walked into Vík to mail them. Passing the usual little concrete houses with painted metal roofs, I noticed gardens, and an asphalted road. Vík was a town of about four hundred people.

At the bank we cashed a traveler's check, before mailing our letters at the post office. Torrential rain began to drum down while we shopped for food. Vík has the reputation of being the wettest place in Iceland. Resigned to a drenching I realized once outside that it was just a shower, intense but short-lived, and ducked back into the doorway until the rain had passed. As soon as the rain eased, we walked back amid the fresh aromas of dampened earth and vegetation.

Close to where we had landed, huge boulders formed a cave-like space. This dry place, requisitioned as a kitchen, meant we had no need to pitch the tent until later, and could watch the evening's antics from there. There was a boat preparing to launch through the surf in an unusual way. Held in place by hydraulic supports, this fiberglass boat with cuddy and large outboard motor looked precariously perched at the end of an extremely long trailer. The other end of the trailer hitched to a huge truck, which had fat tires for the sand.

The whole contraption must have been more than one hundred feet long, yet when the truck began reversing the trailer down the

Vík

beach, it looked too short. With the boat crew on board, the trailer pushed out through the shore-break, twisting and shuddering, bursting through each oncoming line of surf. When far enough from shore, the boat engine roared into action and the supports swung aside, freeing the boat. It lurched out toward deeper water before turning west and speeding away, leaving the wallowing end of the trailer supported by large round floats. The truck, in low gear, steadily hauled the long contraption back ashore and up the beach.

Meanwhile, the surf continued to roll toward shore, where it reared up into a final heaving, dumping, break which pounded the beach. Sand, dragged up from the bottom by each explosion, darkened the white foam. When the swash raced up the steep bank of beach, the brilliant white frenzy spread across the jet-black wet sand. Eventually, the landward probing edges of water began sinking into the sand, while tendrils of fizzing and bubbling froth still crept forward.

Now, the main body of water began to retreat, clawing back into the backwash, building strength, sucking noisily under, and into the next gathering dumper as it reared up ready to pitch forward. It was like Brighton beach in a storm, without the flints, and with the waves on steroids. The air trembled with the shudder of breakers.

Now the boat was out of view; the show over, I decided to make scones. My early attempts using plain flour had been pitiful. Unable to describe baking powder or self-rising flour, and seeing nothing recognizable on the shelves, I had tried to make do without. Those scones turned out like bricks. Finding baking soda at the store here made all the difference; my scones rose.

Searching for the right type of flour was not our only challenge. Finding neither potatoes, nor fresh fish, we came away with canned fish and meat, rice, and porridge oats. There was little on offer in the way of vegetables besides onions and huge

Iceland by Kayak

cabbages. All the package sizes were much bigger than in English shops. There were no small bags of flour, sugar, or rice, and the smallest bucket of jam available barely fit through a kayak hatch.

Such bulky staples left scant room for extras in the kayaks. There was no point in jettisoning basic food to make room for variety, but I could free up space by carrying less gear. So, the next day I packaged up everything I thought I could manage without, including my down jacket, mailed it home and bought a loaf of bread. I looked forward to being able to fit it in my kayak and still have a little free space.

While in town I met an American couple taking a gap year to travel, currently hitch-hiking around Iceland on their way homeward from the Himalayas. I left them shopping and trudged back wearily along the strand. My legs ached. Walking on soft sand took twice the effort.

Immediately above our camp, sheltering us, were the massive rocks fallen from the cliffs, amid banks of earth, all covered in vegetation. There were burrows everywhere. The air vibrated with the sound of grating, grunting, and growling, gentle rumblings, buoyant cries, and the scratching rustle of feathers. I could smell the birds, and the green dampness of plants and moist earth.

Puffins scampered from their burrow entrances looking bewildered, as if caught in the spotlight. Quickly checking all around for danger, they scurried anxiously along the rocks to find somewhere suitable to launch themselves into the air. Delightfully amusing to watch, they flapped their wings in a frenzy to stay aloft as if a moment's pause would cause them to fall. They are not relaxed gliders like fulmars.

Reynisdrangar, the dark weirdly contorted sea stacks where Geoff had recognized Neptune in the shapes, just west of Vík and beyond the end of the hill Reynisfjall, from here looked like witches cast in stone. I could see their long-pointed hats, warty skin, and weathered faces. One legend says these stacks originated

Vík

when daylight caught two trolls dragging a three-masted ship toward shore and turned them into stone.

Three miles or so beyond these stacks we reached Dyrhólaey, *door-hole island*. Sand and alluvial deposits tethered this island to the coast, as they did Ingólfshöfði. A steep, four-hundred-foot-high, slab of cliff with archways and undercut faces formed a large overhanging bowl of a cove. Inside, the swell boomed and echoed as the ocean steadily thrust up and subsided, like a gigantic chest breathing. Seabirds clung to the cliffs and darted everywhere. Geoff was so impressed by the drama of the place that he braved the considerable shore-break to land, so he could film me with his cine camera as I paddled beneath the cliffs. It was a location to savor.

On our leaving, the tailwind grew strong enough to significantly boost our speed. It was easy to track our progress against the changing landscape of mountains close alongside, and this buoyed our spirits. Before, passing the unchanging scenery of the sands, it had been visually challenging to differentiate between the gain of a single mile and ten. With little visual indication of progress for all our effort, we had learned to derive satisfaction from the passage of time measured by our wristwatches instead.

With no shortage of landmarks, we saw the oddest thing. It was a large airplane parked on the black beach at Sólheimasandur. It looked as if it must have just landed there on the sand, for there was no airstrip. It was the strangest sight, miles from anywhere. Long after passing it, we kept looking back, puzzled. Much later we learned it was an American Navy Douglas Dakota. A Super DC-3, or officially a Douglas C-117, this was one of just four such prop planes stationed at Keflavík. These aircraft carried supplies to Höfn, for the Stokksnes radar station.

This one, on its way back from Höfn, in November 1973, was flying in a storm when it suddenly ran out of fuel. Either that or the pilot had switched over to the wrong fuel tank, an empty one.

Iceland by Kayak

In fog, thick enough to hide the wingtips, the pilot sent out a distress call and turned south, away from the mountains, hoping to try an emergency landing on the sea. Dropping below the clouds just above the sand, they instead managed to land on the beach, at the water's edge. Everybody escaped unhurt, although the plane was damaged. When the US Navy rescued the crew, they decided it would be too difficult to recover the aircraft, so they stripped out the valuable parts and abandoned the plane.

Having passed the plane, we reached a foul-smelling river spewing through a gap in the black sands. It smelled strongly of rotten eggs, hydrogen sulfide. This was the river Jökulsá. *Jökulsá* means glacial river, and there are other rivers known by that same name in Iceland, including the country's shortest river which we had already seen. In AD 965, administrators divided Iceland into four parts, or farthings. The Homann Heirs Map of Iceland drawn in 1761 shows the smelly Jökulsá as part of the boundary between the east and south farthings, the latter being only the southwest.

This Jökulsá also gets a mention in the 13th Century Landnámabók, *Book of Settlements of Iceland*. According to this, a man, Loðmundur, having sailed to Iceland, threw overboard his high seat posts to see where they would drift. These were the wooden pillars that normally stood behind the seat of a chief, at home. He settled the land east of Jökulsá where they washed ashore, calling his farm Sólheimar. People know the river as *Jökulsá in Sólheimasandur,* to distinguish it from other Jökulsá rivers.

Loðmundur, a powerful sorcerer, was growing old when Þrasi Þórólfsson, another mighty sorcerer, claimed land to the west of the river. The story tells of a massive flood, and how Þrasi used his sorcery to divert the flood to the east of Sólheimar. Loðmundur, in return, told his slave to hold his staff in the water for him. He wrapped both hands around the staff and bit it, until the flood diverted to flow west of Þrasi's land. The adversaries

Vík

kept moving the flood, east and west, until they agreed on a route for the river.

This river is meltwater from Sólheimajökull, a glacier that descends from Mýrdalsjökull. The caldera of the active volcano, Katla, lies beneath the ice of Mýrdalsjökull. Erupting every forty to eighty years or so, it sends floods across the south coast sands. The flood story told in Landnámabók must have been based on an eruption under Mýrdalsjökull not long after the first settlement of Iceland. The better documented jökulhlaup, of 1918, carried so much debris it extended the coastal plain more than three miles seaward.

Scooting along with the wind at our back, I knew we were making easy miles. But the wind was picking up. There was a weather change in the air, so what was coming?

BOAT LAUNCH, VÍK Í MÝRDAL.

Reynisdrangar Sea Stacks, Vík í Mýrdal.

Vík í Mýrdal.

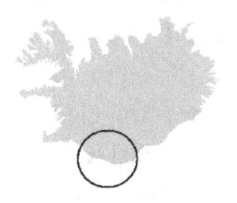

15
Breezy Again

The seas gradually built as the wind increased solidly behind us. Geoff surfed by me, hooting with laughter as he raced off on a steepening crest. Cold, I pulled my anorak hood over my head and cinched it against the rain that began to lash. When I looked for Geoff, he was a full hundred yards ahead. I tried to catch a wave, surfed for a moment, stalled, and picked up the next. Every surge forward was short and unsatisfactory, while Geoff sped ever farther away. I paused, watching the backs of the waves sweeping past until I saw them grow, each standing slightly taller and steeper than the last. This looked more promising.

Finally, my kayak took off. Running from one wave, to the next, and the next, I never fully lost my surfing speed until I had shot past Geoff. The game was on. With each set of steep waves, one or both of us would take off on long swooping rides, urging our kayaks onward for as long as possible before stalling.

Neither of us went far during a lull. It was only when the next set reared up behind us that we raced off again. It was invigorating and rewarding. We were making great progress with little pain, and it was fun. But as the wind continued to freshen, and rain pelted us, I began to grow uneasy. Eventually, surfing up close to Geoff, I called, "It's getting breezy!"

"You're not kidding!" he shouted back, our kayaks lurching as we struggled to keep close without colliding.

"What do you think the surf will be like when we try to land?"

"I don't know. I think we should quit before too long. It will only get bigger." But, by squeezing in one last ride, and then just another, we continued.

Although thrilled, scooting downwind and taking advantage of the following seas, I grew increasingly anxious. I felt torn between the ease of our rapid progress and the growing probability of a scary landing if we left it too late. The shore all along was steep sand, with no shelter from the full force of the weather. Sure, there would be surf breaks to negotiate on the way in, but the real danger was in the size and power of the dumping shore-break.

Indulging ourselves a little longer than we should have, we eventually reined ourselves in and turned toward shore. The moment the steep ominous grey walls of water began rearing up behind us, we knew we had reached the break zone, but the waves collapsed immediately. As the churning water thundered down throwing my kayak shoreward, I struggled in the foam to get any glimpse of what lay in my path. I clung on desperately, blind to where I was heading until the wave threw me aside. I had but moments to get my bearings before the next breaker pummeled me. Carried at last into the punishing shore-break, I hurriedly ejected from my cockpit. Each powerful backwash dragged at my legs as I fought to haul the kayak up the steep bank. Along the dark beach, Geoff engaged in the same struggle. We were fortunate to have stopped when we did.

Wind-driven spray stung my face like gravel. There was no time to waste. Quickly scouting, we saw no possibility of shelter whatsoever. Sand and spray stormed along the beach, and over the berm, splattering across a sheet of water that lay on the other side.

Breezy Again

We had landed on a sandbar between the ocean and a lagoon. The juddering spiky grasses, the driftwood logs atop the bank, and even the rain, were insufficient to stop the wind from stealing the sand. Sand and spume poured along the beach in the squalls, darkening the air. The wind, the rain, and the boom of breakers dizzied my ears.

"This isn't good," I shouted, disheartened. "But what are our options? I'm not launching again. Can you give me a hand to get the tent up?"

Between us we shook out the nested pole sections and grittily assembled the four poles, fitting them into the plastic A-frames and attaching the ridge pole. The wind threatened to take it, so I clung onto one end while Geoff steadied the other. Together we tried to unroll the tent and wrestle it over. A rain squall pelted us, and the tent buckled as one of the pole sections flew apart.

We began again, grappling with the writhing fabric, stretching it over the poles. Geoff managed to roll a log across one corner to hold it down. I clung to the other end until more chunks of driftwood anchored the valences all around the base of the tent. "Is that enough to hold it?"

"I doubt it. We need something heavier. Here, let's try this. Can you give me a hand?" I glanced at the tent, wondering if it would stay there if I let go. It was shuddering and shaking in the blasting wind, running wet. If it flew, we might never catch it. But it held.

Finished with the tent for the moment, I shivered. The wind whipped through my fleece leggings making my legs ache like cracked bones. With every movement, my skin rasped against the grit embedded in the saturated fleece. To change from my wet clothes, I must crawl into the tent which, streaming wet and streaked grey with sand and spume, flapped so much I was afraid to unzip the entrance. I knew it would be cruel inside, the wet nylon whipping my bare back as I changed, so I stalled.

"Let's bring the kayaks up here first. If the wind is like this tomorrow, we might be able to launch into the lagoon." We carried them one at a time to the tent, aligning them as a partial wind break.

"I'm going to get a brew going," I shivered. "I need to warm up before I change."

We both crouched inside the tent, with the door zipped tightly shut, and fired up the stove. Dark smoke shuddered like a living thing between us as the gusts pummeled the tent.

When the water boiled, I stirred extra sugar into the cocoa powder to make sweet chocolate drinks. Then, rather than shut off the stove, we began cooking rice to go with leftover fish from last night's meal.

The roar of the stove, and the thrashing of the tent, was not enough to mask the deep thunder of the surf, yet above all the noise I suddenly became aware of voices, deep and gruff. I opened the tent to find two men, in thigh-length waders, and tough raincoats that gleamed wet. They seemed anxious, uncomfortable about us being there, which puzzled me. Why would anyone mind us camping, especially given the harsh weather?

Although neither man spoke English, their gestures conveyed that it was windy. I nodded, upturned the palms of my hands, and shrugged my shoulders. I stuck up my thumbs and shrugged again. "Yes, windy. Not the best weather, is it? But hey..."

One of the men grabbed the top of the tent and shook it, making me fear for the burning stove inside. He swept his hand above the ridge to simulate the wind carrying it away. He shook his head, pursed his lips, and blew. To reassure, I gestured to the logs holding the tent, and again shrugged to show I saw no problem.

He turned to point, and I followed his eye. Across the lagoon stood a Land Rover, on the grass. Was that a more secure place to camp? It certainly looked nicer, but it was every bit as windswept,

Breezy Again

if not more, and with no driftwood or rocks to use as anchors. When he started to roll aside the logs holding the tent, we had to stop him, bringing him to the front, pointing to our stove inside.

Geoff and I discussed our options. "If they want us to go with them," said Geoff, "why don't we play along and see what happens? It can't be worse than camping here, and it might be better."

Desperately uncomfortable out in the weather and loathe to move, I saw how little time it took the four of us to gather everything. We stuffed the collapsed tent, along with loose items like the already cooled stove, into the cockpits of the kayaks. Wading the lagoon, we floated the kayaks between us toward the Land Rover where a young man stood sheltering beside. He spoke a little English and told us not to put up the tent. One of the men, Karl, he said, had invited us to his house. Leave the kayaks. He would bring a tractor and trailer to fetch them.

We piled into the car and crossed the fields to a farm about a mile away. There, out of the wind at last, we changed from our wet clothes. Welcomed into the kitchen for a hot meal, I sat wearily, my skin stinging in the warmth. Karl, who spoke almost no English, waited until we had finished eating, and then beckoned us upstairs to show us beds where we could sleep. Even from inside we could hear the wind howling. We little knew how strong it would get in the next five days.

I awoke the next morning to the sound of a phone ringing. It rang for a long time unanswered. Too hot, I went downstairs. There was nobody there, but the wind outside was formidable. It blew so hard it made the walls of the house hum. Was that something reinforced concrete always did in wind, I wondered? The experience was so novel to me I could not keep from holding my fingertips against the walls to feel the vibration.

Settling into a chair in the kitchen, I was slumped asleep when Karl came in from milking his cows. His lean face carried a slightly shocked expression under dark tousled hair. He bustled around the kitchen making porridge, laying out bread and cheese and carrying a giant jug of milk to the table.

I tried to ask for a weather forecast, miming and gesturing, and in the end concluded from his gestures that it would be on the radio later. In any case it was obviously too windy for us to leave just yet.

Two youngsters from Reykjavík, staying on the farm while school closed for the four-month summer break, joined us in the kitchen. One of them, Jessie, had an almost permanent smile across his wide face, giving him a cheeky hamster expression. He spoke a little English and translated. "Karl will drive. He asks, would you like to go with him?"

"Of course, yes please, we'd love to!" So, after breakfast, we piled into Karl's battered red, Russian, *Lada* sedan. The car stood low above the road and was badly dented. I could easily see why when we started. The road was rough. Stones flew from the wheels and slammed the metal beneath our feet. Sometimes a wheel dropped into an extra deep pothole, and the chassis scraped.

Karl seemed oblivious to the quality of the road, and uncaring if he might damage the car. Instead, he drove fast, weaving fully from one side of the road to the other, picking the line of least resistance over the corrugated and pitted gravel surface. It reminded me of whitewater kayak racing, where the goal is to sprint down rapids as fast as possible. It is essential to visually plan the best route between the rocks, with the fullest flow of water, as far ahead as possible.

Here on the road, there was no natural flow of water to define a route and no soft landing in the holes. When there was no obvious clear route, Karl steered straight at potholes at back-jarring speed. I was relieved when we pulled up at a building site. With dust billowing around my ears, I climbed out, thoroughly shaken and with my hearing dulled.

Breezy Again

Karl eagerly introduced us to a man who, as a merchant marine engineer, had traveled widely. He spoke English. "We are building a community center for the two-fifty, or so, farm people in the area," he explained. "We lay the concrete foundation, and cast the concrete walls, reinforcing them where needed. The roof will be metal, corrugated. That's the way we build here."

I thought about the little farms we had passed, with often bare cement walls, and a roof brightly painted in red, blue, or green. They stood out well against the drab colors of the landscape.

"Iceland has a lot of earthquakes," he added, "and this construction is earthquake safe." Safe or not, I doubted I would feel an earthquake after our bone-rattling drive.

From the building site, Karl took us to watch the resurfacing of a section of road. Trucks carried volcanic ash from a nearby hillside to spread thickly onto the highway, where a leveler scraped it flat. This ash, he said, makes such a good surface they use it on roads within a radius of fifty miles. Earthquakes, harsh winter conditions, and the small Icelandic population, must make it uneconomical to apply asphalt anywhere except in towns.

Toward evening, the kids at the farm asked if we would like to go swimming. Geoff was keen. I have never been an avid swimmer, preferring to be on the water, rather than in it. Thankfully, nobody seemed in a hurry to leave. As the clock turned its hands closer to midnight, I became ever less enthusiastic, and steadily more lethargic. However, I changed my mind when the call came, loath to miss an outing.

We all squeezed into Karl's car and headed down the road toward the community center. I expected the swimming pool to be in the village, so I was surprised when we took a side turn and raced inland up the mountainside. The road turned into a track so rough we shook like dice, until the car came to a shockingly abrupt halt.

Iceland by Kayak

Piling out to see what had happened we saw how the car, straddling a boulder, had become lodged by its sump. Together we lifted and bounced the car free. Karl waved his hand dismissively and made a walking movement with his fingers. This was as far as the car would carry us.

Following him up the path toward the ice cap, my towel under my arm, everything seemed more than a little bizarre. Although it was still daylight, it was by now early morning and freezing cold. We were climbing higher into the mountains. Swimming up here did not promise to be fun.

On rounding a knoll, I saw a small outdoor swimming pool ahead. Concrete walls formed three sides of the pool, holding water against the rock of the hillside. Hot water fed through pipes down the slope, and in rivulets down the rock surface. The water in the pool steamed. There was a low, flat-roofed building at the far end: the changing rooms.

We changed, and jumped in, luxuriating in the hot water, while the chill wind continued to blow straight down the valley from the Eyjafjallajökull ice cap. The boys, daring each other, vaulted from the warm water, and raced down the hill. After just moments of screaming, writhing in the glacial stream, they sprinted back up the hill and bombed back into the pool.

As the days went by, the winds kept us pinned at the farm, Efsta-Grund. There, seeing our curiosity, everyone took time to show us around. Tucked away in the barn was the small boat Karl used for fishing. We presumed he used it in the lagoon, or during calm days on the ocean. Outside stood a smoke chamber, a rusted steel cylindrical tank, with a capped chimney, and a pipe sloping down to the firebox. Cables tethered it against the wind. "We hang fish inside, sometimes puffins or lamb. People used to hang meat and fish to smoke by the kitchen fire, but…" Jessie's sentence trailed off as he gestured toward the smoke box.

Breezy Again

Jessie was keen to show us fish hanging to dry outside under the eaves and took pleasure in prying off a piece of fish with his penknife and chewing away at it. Karl shook his head dismissively, lifted down a fish and pounded it with a rock to show how to better separate the fibers into something easily chewed. Prepared like that, it tasted good, and was certainly easier to eat. Otherwise, the flesh was as hard as bone.

Karl showed immense pride that evening, bringing to the table something he introduced as *bloomer*s, or at least that is what it sounded like. He wrote down the name for us: *blódmör*, followed by *slátur*. We were none the wiser. It was like a large dark round pudding, cooked in a thin shiny skin. We had no idea what it was.

Suddenly, Geoff said "Ah!" and stood up from the table to mime. He pretended to lift the pudding, tucking it under his arm, and began to walk around as if playing bagpipes, blowing raspberries through his pursed lips to mimic the sound. Everyone collapsed laughing, Karl with tears running down his cheeks.

Geoff had guessed that this was a pudding, like haggis. Karl sliced one apart and pulled aside the sheep's stomach to reveal blood pudding inside.

For other meals, Karl prepared horse-meat sausages, smoked lamb, ham, and a sausage-shaped blood pudding that looked like what I knew in England as black pudding. Using milk from his herd he made soured milk, and skyr, a form of fermented milk I had never come across before. The skyr looked like a slab of lard, white and firm. When mixed with milk, or sour milk, and spoonfuls of sugar, it became a thick creamy, rich, and delicious dessert.

Karl served a choice of breads: white, pale brown, and rich dark brown. At each meal, he set out jugs of creamy milk. Since we sat down to eat six or seven times every day, or so it appeared, the days seemed endless. It was always light, so it was arbitrary choosing when to sleep. With no kayaking to exhaust me, and in the constant light, I did not feel at all tired.

"There ought to be something we can do in return for all they're doing," Geoff suggested when we were alone. So, we approached Karl to offer our help.

Karl discouraged us. Reluctant to put us to work, he eventually showed us a barn, in which remnants of last year's hay and rubbish lay scattered. We could help empty the barn for the new harvest. The two boys joined us, and as we worked, I taught Jessie how to wolf-whistle between two fingers. Struggling at first, he soon got the hang of it. In no time he began whistling so loudly, and so often, I wished I had never shown him how.

Our job completed; Geoff offered to finish mowing a field of grass nearby, already partly cut. The weather was clear, with no rain forecast, so Karl led him to the tractor. Geoff was in heaven. He loved tractors, and this one was small, with no cab. He climbed onto the metal seat and straightened his shoulders with a broad smile. Having followed the tractor on foot, to where Geoff stopped on the field, I confided, "This doesn't look much like a hayfield, does it? It's hardly ready to cut; you could still play football on it!"

Geoff said nothing. Grinning down at me, he shook and wiggled the gearstick until it engaged, and let out the clutch. He would be happy for a while, spiraling around and around the field.

We were at the kitchen table later, after Geoff had finished, when Karl leaped to his feet and ran to the window. Something had just blown past. The wind had changed direction and was funneling violently. An oil drum blew past the house, clomping loudly as it bounced, followed by wood, plastic rubbish, and sand. We watched the action from the window as the hay Geoff cut began taking to the air in drifts, piling up against the fence at the end of the field. It was not long before the fence gave way under

Breezy Again

the strain, twisting and collapsing onto the ground, section by section. "Oops," said Geoff, "that was bad timing."

When the wind showed no sign of easing, Geoff and I started another project, in the shelter of the workshop where Karl worked on his car and tractor. Geoff sketched on a sheet of paper the plan for a pit, for Karl to work beneath his vehicles without having to lie down. Seeing how excited Karl was by the idea, we set to work digging. We labored until we had a pit about eight feet long, four feet wide and five feet deep, and had boarded it ready to concrete.

One evening, Robin drove us to Vík; a rattling road trip of thirty miles to get there. The occasion, early July, was the start of the summer holidays for Icelanders, when everyone tries to get away into the countryside. To further encourage people to spend time outdoors, television broadcasting would stop for the next month. This weekend was a time to party.

People gathered at Vík, the largest town nearby, where there would be a dance. Confronted by a pricey door fee, and seeing little action in the hall anyway, Geoff and I stayed outside with Robin. Robin had installed a quadraphonic sound system in his car, with a cassette deck to play music, so we got back into the car with him.

"This is really good, listen to this!" he boasted, turning the volume up until the car shook. "It plays great!" He steered the car gently along the asphalt-covered street. "It doesn't play very well on most of the roads," he admitted. "They're too bumpy for a cassette player." That explained why he had not played any music on the journey. He was in heaven, following the other cars, mostly enormous American sedans, cruising slowly around and around the few streets of the tiny town. Everyone played music, drank, and called each other by radio as they drove.

I was becoming used to the adventure of life on land here, but I realized the weather must change eventually. We should surely be able to paddle again before too long.

KARL WITH OUR KAYAKS AT EFSTA-GRUND.

EFSTA-GRUND FARM.

16

Vestmannaeyjar

It was the fourth of July when the wind finally eased. "Take a look at this, Geoff." I handed him Karl's binoculars. "The sea still looks lumpy, but the wind has dropped right away. We should be okay, shouldn't we?"

Geoff studied the view. "Yeah, let's give it a go." But at that moment it began to rain. Ordinarily, that would not have stopped us, but Karl's son had baled hay just the previous day and the bales still lay scattered across the open field. "Quick, let's get them in before the rain ruins them."

The lightweight bales were not tightly bound, so we could easily hook our fingers under the baling twine to lift them without cutting our fingers. We followed the tractor around the field, heaving each bale onto the slow-moving flat-bed trailer, before stacking the hay neatly in the barn. The trailer emptied, we lifted our kayaks aboard, towed them to the edge of the lagoon, and waded across to assess the surf where the lagoon emptied into the ocean. Satisfied, we collected the kayaks, finding Karl waiting to see us safely through.

I paused at the narrows for a lull, and sprinted. Only then did I realize how badly I had underestimated the inrushing tide. To my dismay, I was still well within the break line when the first wave of the new set reared up and collapsed. I piled right under it, surfacing just in time to pound into the next crest. I was utterly out of synch and made heavy work of getting through the breakers. When Geoff joined me beyond the break, we waved our final goodbyes and set off toward Vestmannaeyjar, the *Westman Islands*, which stood tall and steep in the mist.

After six land-bound days, I smiled happily at the lively motion of my kayak as we sped along. Yet I could not help thinking back fondly over our time at Efsta-Grund, since that first night plucked from the beach. The storm might so easily have blown us away that night, or later, had it not been for Karl's kindness. We had been fortunate and were grateful to him for looking after us so well. Our unexpected stay on the farm had turned into a precious experience.

Now, we began to see small cream-colored jellyfish in the water around us, while low above the water flew thousands of small black and white birds. These tiny birds zipped rapidly across the waves. "Do you know what they are?" I asked Geoff.

"They're storm petrels," he explained. "The sailors used to call them Mother Carey's chickens."

"Not a lot to eat on a chicken that size," I said scornfully. "They're as small as sparrows! So, who is Mother Carey?"

"The Virgin Mary, Mata Cara, and she's supposed to protect them. You know, helping little birds like that survive out at sea. The sailors often saw them ahead of storms, so they called them storm petrels. They are so small, they figured someone must be looking after them."

"Great," I said. "Just what we need, another storm."

It did not take long to paddle the ten or so miles to the islands, heading toward the gap between the closest, Elliðaey and Bjarnarey. There in the chunky water of a tide race, the extra

Vestmannaeyjar

bouncing made me uncomfortable. "I'm really busting for a pee," I called.

"Me too," Geoff admitted. "I drank far too much milk and coffee earlier. It was too good to leave. Let's find somewhere to land."

Skirting Bjarnarey, at first appearances bound entirely by cliffs, we reached a cave with a hole right through to the other side of the island. It made a broad archway, blocked by five or six massive chunks of fallen rock, with a steep beach of rough boulders flanking the entrance.

We approached tentatively. The air smelled rancid, like sour milk and damp bird lime, the cliffs echoing the hollow sound of water hitting the rocks and the screams of sea birds. The breeze funneling through the archway only accentuated the dank chill. As the water surged and fell, I gingerly wriggled from my floating kayak, steadying myself with one hand on a protruding dome of slimy rock. These rounded rocks proved perilously slippery underfoot as we lifted the kayaks ashore, taking our time, deliberately probing, sliding each foot until it wedged firmly, if uncomfortably, deep between the slick stones.

Geoff, pointing up into the archway, suggested "Let's try to see the volcano on Heimaey from over there." Despite scrambling through the island, climbing higher, the cliff still blocked our view. Only from the kayaks did the vista reveal that mass of wild, rugged, steaming land, the black and rust-red cone thrust up above a jumble of bare ash and lava.

"It looks as if it's burning!" I blurted. "Is it still erupting?" Drifts of white smoke, or vapor, poured down the slope, clinging to the ground like valley mist on a frosty morning. Excited I called, "Let's head for the harbor."

The harbor entrance bore little resemblance to what my map portrayed. My sea chart was incorrect too. Neither of these maps had been redrawn in the four years since the volcanic eruption.

Iceland by Kayak

We could see how much narrower the harbor entrance had become, compared to the old coastline.

The entrance, leading toward Vestmannaeyjar harbor, pinched between a steep, green-capped, dark-peach cliff to our right, liberally spattered with bird lime from nesting seabirds, and a slope of virgin black volcanic material to the left. This new lava and ash, from Eldfell, had threatened the natural harbor that made settlement here attractive. The remaining gap, now constricted to less than three hundred yards wide, inadvertently better protected the inner harbor. Had the eruption not stopped when it did, it could have closed the harbor completely.

Entering the narrows with the town in view, we saw a small almost black beach on our left. There, the dark slope of lava enclosing the beach would screen us from town, ensuring a little privacy. Excepting that every vessel entering or leaving harbor must pass close by, the beach looked perfect; secluded, yet within easy walking distance of town.

The water here being quiet, reflective, Geoff thought it a promising place to fish. "I'll give it a try for a few minutes," he announced. I paddled ahead, beached on the black cinders, and set up the tent, while Geoff drifted close to the rocks, working his line. This time he was successful, bringing ashore four small fish for a tasty supper. Both tired, with our paddling clothes spread to dry, we crawled into the tent for the night.

"Phew!" I exclaimed in surprise. "That's weird. It's warm in here. Is all this heat coming from underneath, from the volcano?"

"I suppose so. It must be a hot beach!" If unsettling, it did feel luxurious to have under-floor heating in the tent.

Kirkjufell, later renamed Eldfell, erupted in January 1973. Despite its proximity to the Atlantic Ocean, this beach still radiated heat. Considering how harsh Icelandic winters can be, I was surprised it had not already cooled. Or was the heat from new

Vestmannaeyjar

magma bubbling up beneath us, rather than from still-cooling rock? Could another eruption be imminent?

The eruption on Heimaey began when a fissure opened less than a mile from the town center. A two-mile-long chasm split open from shore to shore, isolating this corner of the island. A fountain of lava, up to four hundred feet high, sprayed up from the fissure, spreading into a long curtain of fire.

The activity gradually focused around one vent. In just two days, a cinder cone grew around it to more than three hundred feet high, close to half its present height. Lava then overflowed the lip and spread north and east, layering to more than three hundred feet thick in places. A column of ash rose to 30,000 feet. Activity continued until July, by which time lava had enveloped part of the town, with other buildings crushed by the weight of fallen tephra or set alight by fiery volcanic bombs.

Camping on the edge of all that recent action, I had to reassure myself that further activity so soon was unlikely. Despite that, I knew it could happen. I drifted uneasily into sleep.

The next thing I was aware of was the tent door unzipping and a cheery face peering in. I struggled to understand what was going on. Very warm, cozy, and sleepy, almost drugged, I propped myself up on my elbow in my cocoon, my sleeping bag still sealed around the neck.

"Good morning! My name is Sigurgeir Jonasson, and I'm the, um," and he paused, "do you mind if I take a photo?" He raised a large camera to his eye as I squinted drowsily, trying to rationalize the appearance of somebody strange in our tent, and that fat lens aimed at me, just five feet away in the narrow space. What was going on?

"Click. Click-click-click," the shutter repeated rapid-fire. "Yes, I'm a reporter for the newspaper. I've been following your news along the coast. Welcome to Vestmannaeyjar." An interesting welcome indeed, I thought sleepily.

Iceland by Kayak

I crawled from my sleeping bag, and then from the warm tent, onto the crispy cinder beach. It was a bright day.

"Do you mind if I photograph your kayaks too, and the tent?" It was a rhetorical question, and his enthusiasm was pressing. "You know, you are the first to camp on the new land from the volcano," he added.

We posed beside our gear while he shot away with his camera. Then, as abruptly as he had appeared, Sigurgeir hastened to leave. Enthusiastic, animated, energetic, he was also perhaps on a tight schedule.

"You must come and find me in the town when you're ready, any time. Here's my number. Don't be shy. I'll help you if I can."

He passed me a slip of paper with his phone number, then turned, his footsteps crunching loudly across the cinder beach. When he reached the end, he clambered rapidly up a tongue of steely lava, vanishing over the top in the direction of town.

I stood staring blankly after him, then down at the four-digit number he had written. Only four digits? Would there be a phone box at Vestmannaeyjabær post office?

Saving a town visit for later, we took to the hills. Eldfell was calling us, and the top of our small beach clung to her skirts. Her lava and ash were warm and steaming. In places the ground was hot to touch. From within the deeper cracks, the rock glowed red.

Partway up the hill, workers were busy burying huge steel pipes under the tephra. Steam issued from the ends of the pipes. I asked what was going on. "They're for heating the houses in town, a new idea, the first experiment of its kind. It will work like the heating system in a house with a boiler and radiators. The hot water will cool down as it warms the houses. Pumps will push it back up through these pipes, to heat up again under the hot ash. There is more than enough energy coming from the mountain!"

They aimed us in the right direction for the summit, and as we climbed, more of the island came into view. Spread below us, the

Vestmannaeyjar

frozen river of lava led down into the town. Part of Vestmannaeyjabær lay buried under that tongue of jagged rock, where lava had cooled and solidified. Buildings protruded from the cliff, as if the rock had frozen in the middle of swallowing them. Ash piled deep over others. Beyond that twisted edge spread what Eldfell had spared of the colorful town.

Pushing up the steep cinder slope, past an area newly planted with clumps of stabilizing grass, we climbed into mist. Red volcanic bombs, now cold, lay splattered on the dark ash. I pictured how they would have been born, as globs of molten rock, spat upward and outward from the vent, hurtling down in fiery fragments to clatter onto the growing cone of tephra.

Probing, looking for the easiest way up, we reached a rust-red ridge; a natural route to guide us. Steaming patches of brilliant yellow sulfur crystals filled the air around them with choking acidic fumes, reminding me of chemistry lessons at school. The yellow contrasted with the vibrant red rock alongside, and patches of white crystals. Heat blasted from every crack and cleft in the crusty surface. When I paused to look down into a glowing crevice, I saw the soles of my shoes smoking, the rubber melting.

Climbing higher, we saw the rim of the crater materialize from the mist as a skyline edge above us. The ash cone sloped away so steeply behind, our every footstep dislodged clinkers sending them tumbling, tinkling, and clattering downhill out of sight. I picked up a black cinder to inspect it. Tiny bubble-holes pitted its surface, the bubbles broken open to leave fine, knife-sharp blades of glass-like stone between the cavities. The inner surfaces of the holes gleamed iridescent blue and green, vitrified. Examined closely, this piece of ash, just one of millions of fragments making up this dark hill-cone, looked precious. But, sharp-edged and abrasive, it was a beauty to handle with care.

I recalled watching news of Eldfell's eruption on TV, the fiery lava flowing toward the town, a line of dancing flames, and the night sky lit by the flying streaks of lava bombs above the growing volcano. I remembered how vulnerable the houses appeared while, towering above them, a glowing river of lava, edged in fire, crept inexorably toward them.

The media covered the story well. Everyone, except one man, evacuated safely. I became fascinated when teams returned with pumps to spray seawater at the lava, hoping to cool it enough to create a barrier to divert the flow. Their hope was to deflect the lava from the town in its path. The world watched breathlessly as first the pipes melted in the heat, and then the pumps began to fail.

The bold attempt showed defiance in the face of a mighty and indifferent threat. Did it help? Who knows? It may have saved the harbor. Lava might otherwise have completely blocked it. Had the eruption lasted one more week, one more month, or one more year, how much of the town would have remained? In such an emotional situation, the Icelanders' fight against the volcano was an international inspiration. But this was not the first eruption in Vestmannaeyjar in recent years.

As a kid of ten years old, I remember the news of Surtsey, and of Surtlingur, *Little Surtsey,* which rose from the ocean in clouds of steam, throwing out ash. Those new islands, just twelve miles from Heimaey, appeared in 1963. They continued to grow until the eruption ceased four years later. This instilled in me a fascination for volcanoes, as each spat out tephra, growing day by day, and sending a column of ash high into the atmosphere. Surtsey was named after Surtr, the mythological fire giant. The Icelandic suffix *-ey* means *island,* hence Surtr's Island or Surtsey.

At the very top of Eldfell a crumbling wall of soft red rock gleamed with a crust of sulfur crystals. Every deep crack glowed from the red-hot rock inside. My eyes stung and my throat burned with the chemical vapor. We followed the crater rim, which formed a curve around one side of the vent, to where the now solidified river of lava had spilled from the cauldron toward the town.

The way back, having dropped to the lava, was far from obvious. Retreating uphill from each failed attempt, we finally found a way to the base of the volcano. There, closer to town, people were busy hand-planting little clumps of marram grass in the loose ash, while others sowed grass seed.

Sigurgeir was at his home, located as he had described, right by the volcano. He had built an attractive house in Scandinavian style. The spacious construction glowed inside with plentiful natural light and exposed timber. Welcoming us in, he was keen to show us his photographs, which included stunningly beautiful prints of the eruption and its aftermath. He appeared justly proud of his photographic skills. "Look," he said, pulling out another print, "Eldfell here is so hot, this man can bake his bread in a hollow in the rock."

"Are you serious? Was it really hot enough to do baking?"

"Oh yes, the rock is still extremely hot inside."

Here were pictures he had taken of the bleak Surtsey. While visiting other islands nearby, he had shot crisp close-up pictures of seabirds. Colorful elongated eggs, blue, and green, speckled with brown, lay collected in piles on the grass. These came from guillemots, and razorbills, the shape to prevent the eggs rolling from the nesting ledges. He also had images from the puffin harvest.

"How do you catch puffins?" I asked, recalling Geoff's exploits with the net on Ingólfshöfði. Any tips might be useful.

"A group of us go to one of the islands near Heimaey every year. The hunting season starts in a few days' time, in early July. We stand on top of the cliff with a net with a long handle.

"Old puffins learn to avoid the nets, and we leave the ones carrying fish because they are feeding young. We catch up to five hundred a day of those not yet breeding and keep them in the shade until the boat arrives every few days to collect them.

"At the end of the month there's a huge celebration to mark the end of the hunt. Everyone on Heimaey eats puffins, and gets drunk, for about three days. That's the puffin season!"

I fell silent for a moment, adding in my head, imagining the volume. Would that add up to 15,000 puffins each year, harvested from just one island?

Sigurgeir introduced us to his wife, the golf champion of Iceland, and to their son, working the summer as an electrician on Heimaey. He had brought his own son, their grandson, to visit.

"Did you see how volunteers are planting the ash with grass? It gets very windy here, with fierce storms, especially in winter. In the first days, the ash blew around so hard, it pitted and frosted all the windows. It became impossible to see through the windows, and it stripped all the paint from the buildings. It is very sharp. We had to do something to stop it. The grass helps."

The town was busy with new construction. "Eldfell destroyed about one-fifth of the town," Sigurgeir explained, "so about one thousand people lost their homes. On the positive side, the mountain supplies all the new houses with heat. That includes all the hot water needed for central heating, and for everything else. It's the same in all the new public buildings too."

Later, we walked along the streets into the town center to see the new wall of lava. Houses jutted out, partially embedded in the rock where the lava had abruptly stopped. Charred rooms remained half-filled with ash, and ash lay in drifts everywhere on the streets.

It was difficult to imagine the shock people must have had when the volcano unexpectedly erupted, so close to home, in the middle of a winter night. How could so many people evacuate the island so quickly? All those people with buried houses must have

Vestmannaeyjar

lost everything except the few small items they could easily carry. What did they choose to take with them?

Sigurgeir came next day to see us off. We loaded our kayaks and drifted into that naturally modified harbor entrance. The water slept calm enough to mirror the unworldly landscape of Eldfell, evoking the formidable underworld that had thrown it up. Hidden beneath the islands lay a pool of molten rock, like a potent creature ruminating, waiting.

Behind us in contrast stood the grass-topped cliffs of an earlier age, one that far preceded the arrival of the *west men*. Sigurgeir took more photos as we waved farewell one last time, before turning away.

BEACH AT HARBOR ENTRANCE, HEIMAEY.

BJARNAREY, VESTMANNAEYJAR.

AUTHOR ON HOT TRAIL BETWEEN ASH AND LAVA.

17
Salmon

Leaving Vestmannaeyjar on a calm sea, we crossed to the mainland and followed the low sandy shore northwest. As my kayak cut through the familiar clear dark water of the Atlantic, I suddenly saw a weird change in the color of the water ahead. Across my path lay a border of chalky water, pale blue and cloudy. Was this the outflow of the Þjórsá River? Where the fresh water, cloudy with glacial silt, forged a channel above the denser saltwater it created an uncannily crisp surface boundary line. From this abrupt cloudy surface intrusion hung an opaque twisting curtain of silt, extending far below into the clear Atlantic. I hesitated for a moment, almost fearful to cross this eerie edge, but my kayak ran onward. My bow overran the line and with a jolt my kayak spun, shoved by the powerful current running out to sea.

The settled conditions encouraged us to press on, past sand and lava field. Occasionally, protected passages behind ledges offered respite from the surf. My mind wandered. The rhythm of my paddle strokes marked time to a mantra cycling in my head, and though my eyes remained watchful, my thoughts were elsewhere. I ran on automatic pilot. Hours floated past while my body worked away, disconnected from my thoughts. As we crossed the broad bay toward Þorlákshöfn, far from shore, I stared with detachment at the distant sight of a small, solitary, inflatable boat. It drifted, people fishing.

Wearily entering Þorlákshöfn harbor to look around, and finding nowhere better to land, we settled on a stone slipway. It was nighttime, still light, nobody about. Fishing boats clung to the quay.

"There doesn't seem to be anywhere to camp. Do you want to keep going, or sleep on the quay?"

It was an easy decision to make. After more than forty-five miles, I had little enthusiasm or energy to paddle farther. With no time wasted, the roar of our stove soon closed around us, beside the tent, at the top of the slipway on the stone quay.

We had almost finished eating when two men startled us. "My name's Emil," one introduced himself. "We've just come ashore from fishing. We were catching salmon at the mouth of the river and saw you pass. You might have seen us?"

"In an inflatable? Yes, we did."

"Would you like to come for coffee?" he offered. The other, who's handwritten name looked like Grímur Markýsson, added, "Bring your wet stuff and we'll dry it for you." Scooping up our wet clothing, we jumped into their car for the short drive to where we helped unload their catch of massive silvery salmon.

"Wow, what beautiful fish!" I had to admire the slinky scales, glistening like burnished chain mail. "Yes," they agreed. "We net them at the river mouth. Not really supposed to but it's the simplest way to catch them, and everyone gets a share. Even the police," he added with a conspiratorial wink.

Shortly, I sat sleepily cradling a mug of coffee, wondering if police would turn a blind eye in the UK too. Besides, I mused, would someone netting a river mouth commit a sea, or a river fishing offense?

Salmon

Our conversation waning, coffee cups empty and paddling clothes hanging, it was time to trudge back to the tent under the dull and colorless sky.

In the morning, with our kayaks lying on the flat of the quay beside the tent, it was all too easy to take our time cleaning, drying and stowing items. Our camp lay between the fishing boats and town, within a short walk of each. Since everyone stopped on their way past, to look, we took time to show the kayaks, and to describe our route so far.

CAMPING AT ÞORLÁKSHÖFN.

In this town of no more than five hundred, everyone knew everybody else. Any details shared with one were soon known by the next. When a fisherman, on his way to his boat, invited us to coffee, we left everything laying in the open and boarded his trawler. Understandably, he was keen to talk about fishing, and fish stocks.

181

Iceland by Kayak

"There's been so much overfishing, Iceland limits where and when we can fish. But it's a better situation than before: a wider area under Iceland's control, special places set aside for the fish to breed and restock. That was impossible before when anyone could come to fish here.

"Herring boats used to come from all over before the herring disappeared. There was no conservation, no incentive to save. If we caught less, to save the fish, someone else would catch them instead, so there was no incentive to stop.

"We learned a hard lesson. No herring. Cod could easily go the same way, by overfishing. We can't let that happen. Fishing is essential to Iceland; it is our biggest industry." He took a sip of coffee and continued.

"There are very few harbors on the south coast, none between Höfn," his finger stabbed the chart, "and Vestmannaeyjar, and then only here." His finger hovered over the bay where the Ölfusá River met the ocean, where Þorlákshöfn tucked in behind a blunt point at the western end of the bay.

"Our harbor is large enough to take freighters. These others only have small fishing harbors." He pointed to Stokkseyri and Eyrarbakki, fishing villages we passed last night within ten miles of here.

"Eyrarbakki was the main fishing and trading center for the whole of the south, all the way across as far as Vatnajökull, for hundreds of years. Rowing boats fished from there 'till the late 1800s, but the natural harbor is not so good. It is better here, more protected.

"When Eldfell erupted, nearly everyone evacuated to here, the closest good port." What, I wondered, was the impact of the sudden influx of ten times Þorlákshöfn's total population? He continued before I could ask. "Then, of course, to the west there's Grindavík, a bigger fishing port, twice Þorlákshöfn.

Salmon

"Eyrarbakki?" his finger skipped back across the bay on the chart again, east, and he tapped. "Now, that has interesting history. In the year 985, Bjarni Herjólfsson, a merchant, sailed from there toward Greenland to trade. The wind blew him off course and he discovered North America. When he got back to Greenland, he told Leifur Eiríksson what he had found, so Leifur bought his boat from him and sailed there himself."

Leifur Eiríksson, or *Leif Ericsson* as I learned in history, was the son of Erik Thorvaldsson: *Erik the Red*, who, banished from Iceland, settled in Greenland. I had learned that Leif was the first Norseman to discover North America, more than five hundred years before Columbus. Do we credit the wrong person?

"Did you hear about the crazy people who came here in a leather boat, a *curragh*, from Ireland?"[2] he asked, his voice incredulous. "They stopped in Iceland last winter and left from Reykjavík in May. It was all in the news: they reached Newfoundland, about a week ago. They were following the route of Saint Brendan, an Irish monk they claim visited America five hundred years before the Vikings.

"The Vikings had sunstones and lodestones to help them navigate. Who knows how the Irish got there. Nowadays there's all this," he gestured at the consoles in the wheelhouse. I saw radios, depth sounders and radar.

Returning to his chart, he pointed out what we might expect next. "Now, Reykjanes, here, has strong currents. You had best go around at slack tide. Stay close to shore, here, where the current is weaker."

Returning to our kayaks, Geoff discovered a note from Emil who, having learned last night that Geoff had a leak in his kayak, had left a tube of silicone sealant for him.

We had launched and almost reached open water before I heard a call. Glancing back, I saw someone running along the quay, shouting, waving something in the air. Grímur was out of breath and clearly relieved to see us turn back. He passed gifts down to us: a thick pair of woolen socks each. "From my mother to you. She knitted them." He next handed down a large cold package, heavy in my hand. "This is salmon, milk and butter," he explained. "Cook the salmon in the milk and serve it with the butter."

Setting off again, with warmed hearts, Geoff suggested, "Let's try for Grindavík today," thirty to thirty-five miles away. "Then, we'll be ready to catch the tide around Reykjanes. Once we get to Keflavík, we'll call Salóme and see if she is still willing to pick us up. I quite fancy a hot shower and a sit-down dinner!" He smiled in anticipation. I smiled too. I was content, savoring the comfort of dry clothes.

Close beside us stood low cliffs of contorted lava, while beyond, in the distance, shapes like volcanoes poked up through the mist. One stretch of cliff, bold red and brown, looked like hard-packed volcanic ash. Seabirds nested everywhere. Some five miles short of Grindavík, noticing a beach partly sheltered from the swell by offshore rocks, we chose to land.

Beyond the beach spread a lava field, a wonderfully weird world of spiky rock, huge bell-shaped chambers, and gaping holes. The undercuts and hollows revealed the color of the rock to be dark grey, black, umber, and red. Elsewhere the tiny lichens of green, black, grey, sometimes orange encrusting the exposed rock, blended into grey with distance.

Ducking into a crevice in the tortured rock, we startled a sheep from a small but remarkably level patch of grass and gravel. We followed, climbing to a wider view across miles of barren lava field sparsely dusted grey and green. Up there, scrambling cautiously across the lava, I stopped abruptly. There was a hole at my feet. The opening revealed a chamber below, the size of a room, and I peered down as from the lip of a squat bottle. If I dropped inside, I would be unable to get back up the overhang to the exit: it formed a natural oubliette, a deathtrap. As I more warily explored onward, I saw more of these giant bubbles within the rock, suggesting how the molten lava had trapped pockets of gas as it cooled. Inside the chambers, and on the wildly spiky undersides of exposed ledges, twisted fingers of rock hung like volcanic helictites.

"This looks a great spot for the tent," said Geoff, surveying a sheltered hollow surrounded by dark lava. In it grew a small pad of grass sizeable enough for the tent. Looking around I saw sufficient driftwood to build a house.

"Excellent choice," I agreed. This would have made a dream camp for me as a kid. There were natural ramparts to hide behind, nooks and crannies to explore and enough stuff to build a shelter. I was back in childhood again in my imagination. "I could live here!"

"Look," Geoff pointed out the broken shell of a rusted steel oil drum, "we'll use that as an oven."

With a fire lit beneath it, the drum, with a gaping hole in its side, made an excellent stovetop. Through the hole, we set a pot of water to boil. "Now where," Geoff turned excitedly, "where is that salmon?"

Opening the gift bag revealed a stick of butter and a carton of milk. The other cold weight, bound in paper, must be the fish. Unwrapping revealed a huge chunk of pink salmon, in gleaming silvery skin. Magnificent, the skin shone like jewels. It must have come from an enormous fish, its girth too great to encircle with my two hands.

I could only remember eating salmon from a can. Fresh salmon, along with lobster, caviar, and oysters, was too pricey except for on special occasions. But it had not always been expensive in England. Word has it that London students once complained about the dull, repetitive menu of salmon every day, then the cheapest food available. They petitioned for a weekly meal of meat. In those days, the Thames abounded with salmon.

Over time the water became so polluted it died biologically, although *dead* is not entirely true, since it teemed with live pathogens. Anyone falling into the river needed their stomach pumped. Nobody saw salmon in the Thames between 1833, and just three years back, 1974. That landmark moment, in the long Thames clean-up campaign, came just after Norway and Scotland began their experimental salmon farms.

I was glad to have the cooking instructions, to carefully simmer the fish in the milk, and serve it up with the butter melting on top. "Wow, exquisite! I can understand the police turning a blind eye, tempted by this!"

"Mmm, a real treat," agreed Geoff.

The meal having made the evening perfect we lingered longer, feeding sticks to the fire, and luxuriating in the radiant heat well into the stillness of night.

GEOFF, REYKJANES.

Salmon

FISH DRYING. (ABOVE AND BELOW).

Iceland by Kayak

MAP 6. SOUTHWEST ICELAND.

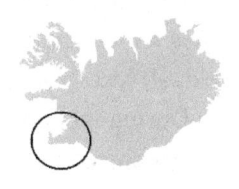

18
Reykjanes

Carrying my breakfast, I tripped and spilled everything before even tasting it. That clumsiness was the harbinger of things to come, for I overbalanced getting afloat and toppled face-first into the water. Doubly discouraged, I emptied my kayak with resentment, soaked to the skin before leaving the beach. Such trivial mishaps should have been forgotten in a moment, laughed away in the next conversation, but my grumpiness persisted. Does a dark mood beget negative consequences? More likely we simply view subsequent experiences in a less positive light.

Reykjanes, subject to strong tide streams, lay just fifteen miles away. By the time we could get there today, it would be too late, the tide unfavorable. But by getting closer we would be in a better position to take advantage of the tide around the peninsula tomorrow.

Reykjanes is one of Iceland's most volcanic and seismically active areas. It lies on the tectonic separation zone between the Eurasian plate and the North American plate, a continuation of the Mid-Atlantic Ridge. This is where new land is growing, where

Iceland by Kayak

Iceland adds one inch of territory every year between continental plates which continue to drift apart. The low promontory is lava field, all less than seven million years old.

During the most recent major volcanic activity, of the 10th to 13th Century, molten lava sprayed from cracks in the rock. Between AD 1210 and AD 1250, lava blanketed an area of twenty square miles (50 km^2). Yet even between the periods of greatest activity, as now, there are still boiling mud pools, hot springs, steam vents, smoke, and frequent earth tremors.

When Vikings settled Iceland, in the late ninth and early tenth Century, it was during a period of eruptions. Undoubtedly all the smoke and steam on this headland made it easy to describe. They called it Reykjanes: *smoking ness*, a ness being a nose or tongue of land.

We began to look for somewhere to camp after passing Grindavík, pausing to consider landing in a bay. At this level of tide, the bay had drained to shallows wherever it had not dried to mud and stone. Two or three houses stood clustered behind the bay toward the eastern end. There would be enough separation between us, for courtesy and privacy, to camp on the western shore.

Deciding to land, we found our approach impeded by a labyrinth of boulders at, or just beneath the surface. If our approach was awkward, similar seaweed-covered rocks, treacherously slick, made landing worse. Carrying the laden kayaks ashore proved an ankle-twisting cruelty, dampening my enthusiasm long before I began looking for a place for the tent.

"There's absolutely nowhere flat enough," I pointed out irritably, searching farther and farther from the kayaks.

"How about here?" suggested Geoff happily, pointing to a place I had already rejected. Between the rocks a serpentine patch of grass, smaller than the tent's footprint, sloped inward between the heads of half-buried boulders. "Look," he kicked aside the

Reykjanes

piles of rotting seaweed and stepped back, startled by the frenzy of sandhoppers erupting from underneath.

He peeled back the encroaching grass from a partly buried boulder, heaved it from its peaty bowl and dropped it aside. Sinking to the ground he curled himself to fit between the larger stones. "It's quite comfortable, really."

I reluctantly tried the adjacent depression for size. "Well, it might work, for one night but it isn't ideal."

Geoff smiled. He always harbored a positive 'make-it-happen' attitude. To park the tent there was easy, but to hold it with stones proved challenging. The frame perched between inward-sloping boulders, so our attempts to pin down the tent valences only partly worked. The rounded rocks, all that was available, either slid or rolled inward. They barely held, even against this light a breeze. Leaving the tent precariously standing, we changed and settled down to snack with a hot drink, which always made a new place feel like home.

"You realize we are out of water?" I pointed out. "We'll need more tonight, and for morning, and I can't see where to get any, can you?" The bay and surrounding shore were boulder-strewn and low. If there was any fresh running water, it was well hidden.

"Let's ask at those houses," Geoff suggested, but the incoming tide, already flooding the bay, made any shortcut impractical. It looked too muddy anyhow.

"We could paddle across when the tide comes in," I suggested, although on second thoughts, why change back into wet clothes, and carry the kayaks again? Besides, it would be too long to wait.

Balking, we chose to walk the circuitous route around the back of the bay instead. Stepping carefully from boulder to boulder, we set off carrying our empty plastic milk containers to fill. On the way, the low cloud began to drizzle on us.

"It doesn't look as if there's anyone here," I noted as we approached. The old buildings looked abandoned. To one side of the main house ran a long, steep mound in the turf. Three chimneys protruded from the grass. It reminded me of a long barrow, a type of Neolithic barrow common in the south of England, except those chambered burial tombs had no chimneys.

At the end of the mound was a wooden wall, with double doors, another sign that this was not just a bank of earth.

Inside, steps led down into a long empty hall, dug down into the ground. It smelled of earth. Pale roots dangled between wooden roof boards. Our voices sounded muffled in the still air. The musty atmosphere felt so comforting and cozy, it invited me to linger. In such peace, I could easily curl up and sleep soundly, but for our quest for water.

When no taps or pumps showed on the outside of any of the buildings, we peered through the windows into a house and saw it was empty. We knocked, just in case, before lifting the wooden latch and swinging the door open.

Inside looked long deserted. Investigating from room to room, clunking across dusty wooden floorboards, did not reveal a faucet anywhere. "They must have had a water supply," said Geoff. "Maybe there's a well. Did you see a pump or anything outside? You would think they would have a tap here somewhere."

I jumped at a sharp rattling sound. It came from above, from the metal roof. Outside fell a deluge of rain. Taken aback, I exclaimed, "Geoff, I didn't bring my jacket." Recognizing he had not brought his either, "We are going to get soaked!"

"You never know," he said, calmly accepting, "it might ease off again. Anyway, let's at least collect some water while we're waiting." Propping our open bottles under the curtain of water from the roof, we resumed our exploration.

"Look, there's one here!" said Geoff in surprise, pulling aside rubbish to reveal a pipe. It stood low to the floor, against the wall in a room which appeared to be neither a kitchen nor a bathroom.

Reykjanes

"What an odd place to have a tap. There's not even a drain." But there was no need to find out if it carried water.

We stood in the doorway to watch the downpour. "You know, if this doesn't ease off, we'll have to go sooner or later," warned Geoff. "I'm hungry."

A wall of water gushed down in front of our eyes, pouring straight off the roof. "It won't be as bad once we're actually out there."

I looked at him astonished. "You've got to be kidding! We'll get soaked in an instant."

In the end we cut our losses and set off, carrying our bottles of rainwater. Chill rivulets poured down my neck against my skin, and my head felt compressed and frozen. My jeans clung heavily, making me walk stiff legged. The tent may have been less than a half-mile away, but I suffered a marathon to reach it, awkwardly waddling from boulder to boulder, wet denim shackling every movement. Adding to my discomfort, I knew how quickly water would have collected in our pans, had we waited at the tent.

The plod back gave us time to plan for tomorrow. "Leaving here at six or seven in the morning should get us to Reykjanes at slack tide. That way, the tide will help us north along the end of the peninsula. It is about thirty-five miles to Keflavík. If the weather's good, we could be there tomorrow. Then, we can call Salóme, tell her we haven't given up yet." Geoff was already there in his imagination. I felt so uncomfortable I could not even imagine a warm shower.

Approaching the tent, I saw my kayaking clothing, spread out to dry, even more sodden. It would be pointless to gather it up, it would still be wet in the morning anyway. I unzipped the tent and crawled in, almost gagging at the rank stench of wet rotten seaweed. There was little space between the sagging sides of the tent, heavy under the rain. A fine mist sprayed through the fabric.

Iceland by Kayak

Too dejected to cook, we turned in. Saturated, I did what I had done before with wet clothes in the mountains; left them on, got into my sleeping bag, and hoped my body heat would be enough to dry them by morning. Once warm, I was tolerably comfortable.

When I awoke, I was lying in water. It was still raining hard. I lifted my head to see water pouring in rivulets from the boulders outside, over the groundsheet. It had pooled in the low spots, exactly where we each lay in our own puddle. Geoff still slept. Realizing how wet everything was: sleeping bag, land clothes, paddling clothes outside, and the tent too, I felt discouraged. Why pack all this wet stuff for a six o'clock start? I decided not to wake Geoff and went back to sleep.

The deluge had eased to a drizzle by the time we crawled from the flooded tent. Hungry, having skipped our evening meal, it was easy to gorge on rice with raisins before cooking scones. There was a strong incentive to keep the stove burning, to warm ourselves with a succession of hot drinks.

Geoff, always planning, began a construction project. He heaved together logs to make three tripods. Under one, he built a huge pile of driftwood, which he soon had blazing. Taking his knife, he cut rope from the remains of a net half-buried in the beach to string between the remaining pair of tripods. Over this line, and across the balanced logs, we threw wet clothes, sleeping bags and tent, even though it continued to rain lightly. Standing close enough to the flames to see the mist rise from the clothes on us, we turned ourselves like a spit-roast whenever the heat grew too intense. There was no way to prevent the flying embers which seared little holes in the hanging nylon items. But it was heartening to watch everything, including the sleeping bags, become dry, albeit smoky, despite the drizzle. I at last began to cheer up.

Reykjanes

With kayaks loaded, and the fire doused, we were ready around six-thirty in the evening. Departing a full twelve hours later than originally planned offered us favorable tides once again,

As we cruised contentedly past low cliffs of contorted lava, the sea grew steadily rougher, bouncier than I expected for slack tide. Gannets cruised by. The light, reflecting from the white of their huge wings, grabbed our attention each time the gliding birds switched direction on their meandering course.

Gannets impress me. Despite their size, carrying a wingspan of six feet, they fold their wings and nosedive from up high to plunge for fish. I can only imagine the impact as they slam into the water. The gannets here may have come from the off-lying rock Eldey, home to one of the world's largest gannet colonies.

Eldey was in sight, eight miles offshore, to the left of the cliffs ahead. It looked solid, a steep chunk of white-topped cliff jutting from the ocean. The 250 feet-tall blocky rock stands above a strange area of shallows and currents which stretches out for forty miles southwest from Reykjanes. Those shallows and tricky seas are known as Fuglasker, *Bird Skerries*. They are dangerous for boats. The sea breaks over the submerged and exposed rocks where, especially during spring tides, the currents cause heavy overfalls.

As the name Eldey, *Fire Island,* alludes, the area is volcanically active. In 1783, an undersea eruption covered the sea here so thickly with pumice that it interfered with shipping. A crater appeared above water, named Nyey, *New Island*, although storms washed it away that same year. The volcanic activity, and earth-shifting, is ongoing. Before 1972, five years back, a rock named Geirfuglasker stood thirty feet above water to the southwest of Eldey. In 1972 it collapsed and became awash when volcanic activity changed the sea depths all around it. That was not long before the fateful January 1973 eruption on nearby Vestmannaeyjar.

Iceland by Kayak

The disappearing rock, Geirfuglasker, was not the only one of that name in Iceland, and not even the first of that name near Eldey. This one took the name of an earlier island here, itself submerged by a volcanic eruption in 1830.

Up until 1830, Geirfuglasker had been a difficult stack for people to land on, so it offered a safe nesting place for sea birds. It was one rock where the flightless great auk, known in Iceland as the geirfugl, *garefowl* or *spearbird*, nested. In fact, it became something of a last haven for the great auk. After the rock submerged in 1830, the great auks that had nested there moved to nearby Eldey. Unfortunately, boats could access Eldey.

Great auks were already rare. Once people reported seeing them on Eldey, in 1835, museums rushed to secure skins for display. Icelandic sailors killed the last two on Eldey in July 1844. By account, these birds were the last pair seen alive anywhere. The last individual killed by man may have been at Vardø, in northern Norway, near the Russian border, in 1848.

In the end, hunters and collectors wiped out these large seabirds from all the remote places in the North Atlantic where they once nested. Three men caught the last great auk seen in Britain, in 1840, on Stac an Armin in the Scottish St Kilda archipelago. They tied it up and kept it for three days. When a storm brewed, the bird became agitated and vocal. Surmising it was a witch in disguise summoning the storm, the men beat it to death with a stick.

Great auks were once a significant, easily caught food source for Native Americans. Later, in the western Atlantic, sailors heading for the Newfoundland Grand Banks saw the swimming birds as a sign they had neared their destination. The last time anyone reported seeing a live great auk was there, in 1852, on the Newfoundland Grand Banks.

Now the great auk is extinct, there will be now more new skins available, ever. That makes existing skins more valued by

Reykjanes

museums. In 1971, the Natural History Museum of Iceland paid a record £9,000 sterling, at that time $21,600 US, for a single stuffed specimen.

Following the gannets, we sped around Reykjanes in the gathering gloom, favored by both breeze and tide. Our swift progress encouraged us to press on for Keflavík, despite the late hour. Anxious to get there, we took a single break ashore, otherwise snacking afloat every two hours.

Swells, breaking across ledges toward the northwest corner of the peninsula, forced us to detour repeatedly into deeper water. Not until clearing the final half-mile-long ledge at Garðskagi could our bows turn toward the southeast. Despite a headwind, the calmer seas and lack of detours made our progress swifter.

In the early morning, fishing boats stood propped up for maintenance or storage at a silent boatyard on the outskirts of Keflavik, with fish factories along the shore ahead. Our kayaks cut through water that thickened, dulled by a surface film.

"This looks like the back door into town," Geoff complained. "It's a mess." The water smelled fishy, and my wet hands slipped on my paddle with the slick oil. Fish scraps floated everywhere. Garbage littered the beaches. "Let's go back to where those boats were. We'll be too far into town if we keep going."

At the boatyard, our kayaks scrunched quietly onto the gravel beside a slipway. Realizing the kayaks would block any other boats landing there, we carried them farther ashore, setting them down between two deep-hulled wooden boats which towered above us, shored up on their keels. There was a strong smell of paint.

When the rain started, seeing no appropriate place to pitch our tent, both of us stood close against a wooden trawler, its hull partly scraped for repainting. Sheltered there from the rain, but not from the breeze, it was time to change from our wet gear. Firing up the stove on the ground amid the flakes of colored paint, I cooked scones and mixed hot chocolate drinks. Meanwhile, the pitiless wind eddied around the hull, reaching us no matter how we tried to evade it.

Yawning, Geoff looked at his watch. "It's a tad too early to phone Salóme, isn't it? But it feels nippy here. Do you fancy walking into town to look for a phone box, for later?"

"Yeah, I'll come. It'll be warmer walking than waiting here. Can we look for somewhere for coffee too?" The thought of a cozy café appealed. Then, I looked around at our scattered gear. Keflavík was by far the biggest town on our route so far. "Do you think our stuff will be okay left like this?"

"Oh, I doubt anyone will mess with it," Geoff said, surveying the kayaks. "I expect only boat people will come here, and they wouldn't touch anything." Stowing our gear against the rain, we sauntered off along the road into Keflavík. In the misting rain, I thought we were the only ones at large at that early hour. We were not.

Reykjanes

LAVA FIELD.

HOME FOR THE NIGHT.

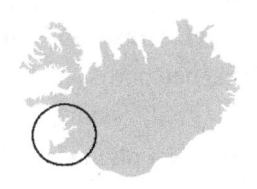

19
Keflavík and Reykjavík

Reykjavík, although small for a capital city, dwarfs the village-sized settlements along the south coast. In 1973, when Iceland's population numbered 203,000, more than 80,000 lived in Reykjavík. Since the adjacent towns, Hafnarfjörður and Kópavogur, housed another 10,000 and 11,000 respectively, all but half the residents of Iceland lived in this one metropolitan area in the southwest.

Reykjavík developed where Ingólfr Arnarson made his home, having moved from his first landing place at Ingólfshöfði in about the year 874. Credited as Iceland's first Norse permanent settler, records suggest that Irish people lived there before that date.

If Keflavík, twenty-five miles from Reykjavík, and a town of 6,000, was worth exploring, it seemed too bleak and empty this early in the day. A dull sky leached the color from wet, red-roofed, buildings and the drab empty spaces between them. Light drizzle, just enough to dampen us as we walked, made everything appear finely granulated. At four in the morning, we were too early to expect to find coffee.

I walked automatically, half-asleep, and barely noticed when a car cruised past in the opposite direction. What alerted me was the sound of abrupt braking, tyres protesting a quick turn in the street. The car that pulled up beside us was a police car. The driver reached across and wound down the window.

"Hi!" said Geoff, bending forward. "Do you know where there's a phone box?"

The stern expression on the face of the police officer screwed into puzzlement. "Where are you from?" he asked. "And why are you out so early?"

"Oh, we're from England," Geoff began, "we're kayaking around…"

The puzzled expression lit into a beaming smile of recognition, as the policeman cut in, "Yes of course! I saw you in the newspaper! Welcome to Keflavík! Did you say you need a phone? You can use one at the police station. I'll take you there.

"I thought you were up to no good when I saw you," he explained as he drove. "There isn't usually anyone on foot at this time of day unless they are drunk. I was sure you two were drunks. The guys at the station will be amazed when they see who I've caught."

In the upstairs office of the police station, welcoming officers pulled chairs to a table for us and poured mugs of coffee. They asked all about our trip, who we had met and where we had been.

When we asked if our kayaks should be safe at the boatyard, they said, "Keflavík doesn't really have any crime. Our main problem is people getting drunk. There is seldom anything more serious than what comes of that. It would be an accident if anything got damaged."

"So, the phone's there," said the officer who brought us here, pointing.

"Oh, thanks. We met a girl on the ferry who said if we called her when we reached Keflavík, she would come and pick us up. She lives in Reykjavík, or Keflavík, one or the other. Do you think it's too early to call?"

Keflavík and Reykjavík

"Oh no, you can't call her this early," someone protested. "You should at least wait until nine. Do you want more coffee?"

"I read about you in the newspaper when you reached Heimaey," someone else began. He was grinning. "There was a photograph of you both lying in your tent in your sleeping bags. It looked as if you had just woken up."

"Or maybe, you hadn't woken up yet," someone else quipped with a laugh.

"That's more like the truth," I admitted ruefully. "I was fast asleep when the photographer unzipped the tent and stuck his head in. I had no idea I was about to have my photo taken, and certainly not for a newspaper."

"Did you see the article?" We had not. "You should be able to get a copy in Reykjavík. It would make a good souvenir for you."

I was buzzing with caffeine by eight-thirty, when our hosts considered it reasonable to call Salóme. Geoff sprang up, "I'll call." He fingered a slip of paper from his jeans back pocket, unfolded it, dialed the number, and listened.

"Hello? Is that Salóme? Yes? This is Geoff. Geoff Hunter?" There was a pause. "Do you remember the two kayakers you met on the ferry to Iceland? I'm one of them: Geoff."

"No, we haven't died yet."

Hearing only Geoff's voice, we had to guess Salóme's reaction at the other end of the line.

"We've reached Keflavík, "continued Geoff.

"Exactly where? Oh, at the police station.

"It's a long story. The police arrested us.

"Yeah, I know, but you know how it is. Can you bail us out? Do you think you could do that?"

All turned quiet in the room, all attention turned toward Geoff. Everyone followed his deception and held their breath. It was too fun a joke to spoil.

"We were drunk. Drunk and disorderly. Look, I know. It is a big favor to ask, but it would really help if you could come and pick us up."

"Yes, drunk. I'm sorry."

"Yes, at the police station in Keflavík. You will? Oh, I can't thank you enough! I really appreciate it. Can you come right away? Well, how long do you think it'll take before you get here?"

When Geoff hung up the phone and turned with a broad smile, the room erupted into applause. Everyone was anxious to get the first sight of Salóme when she arrived, and to see her face the moment she realized Geoff had been joking.

When Salóme parked her red Mercedes outside, every officer's face pressed against the upstairs window. They watched until she entered the building, when they all scrambled to their places, like schoolchildren warned of an approaching teacher.

When she entered the office, everyone sat, stern-faced, official, expectant. She glanced around the room, took in the serious faces and the pair of us at the table, greeted us and in the same breath asked something in Icelandic. For long minutes, all talk was serious. In the end, when someone began to laugh, she realized we had fooled her. She took it well.

Steering for Reykjavík, Salóme sighed, and admitted she had been partying till late, creeping home around five-thirty in the morning, worse for wear. When the phone rang, she had almost ignored it. Even now she was struggling, exhausted and still a little drunk. She laughed. "If you had called earlier, you wouldn't have reached me. I would still have been at the party. Oh, my head hurts!"

Salóme lived with her parents. She introduced us to them and reintroduced her mynah bird. "I didn't have any problem bringing him through customs in the end," she explained as she prepared food. The three of us hungry but tired, after eating we slept the rest of the day.

Keflavík and Reykjavík

The long days of northern summer worked to our advantage. It mattered not which hours we slept. After another meal we went to see a movie with Salóme's friends, Kathy and Bjarny. That film, *Journey into the Unknown*, was the prelude to a car journey into the unknown, into Reykjavík to see the old town. Here, instead of the more usual modern concrete buildings stood stone houses, with corrugated metal cladding over the upper walls and roofs. The city appeared clean, spacious, and, at least at that time of night, un-crowded.

During our Reykjavík stay, Salóme generously took time off work to take us from place to place. She took us to the newspaper office for an interview, where we posed for accompanying photos. There, we picked up a back-copy of the paper, with the photo of us in our tent. She also took us to meet the coastguard, who reprimanded us for not making contact earlier.

When Salóme had an appointment to keep, we mailed postcards, and compared experiences with a Dutchman we met on the ferry from the Faeroe Islands. He was taking his time, traveling around the ring road by bus and hopping off wherever he wanted to explore. He would catch the bus again on another day once he was ready to move on.

Later, when Salóme met Bjarny and us in town, she looked gorgeous. Smartly dressed, she had her hair freshly done. If we appeared slovenly, she did not seem to mind, taking us to the posh part of the city first. There we saw the president's house before scaling the landmark tower of Hallgrímskirkja church, the tallest building in Iceland, to see the view through its narrow windows.

Taking Bjarny and Salóme's father with us, Salóme drove us back to Keflavík, to collect our wet kayaking clothes to launder. Bjarny, intrigued by the kayaks, took mine afloat. Trying to turn back to shore, he capsized, flailing underwater until he remembered to release the spray deck to exit.

Iceland by Kayak

Salóme's father, seeing his panic, quietly took off the spray deck he had borrowed, prudently deciding that this was not the right time to try. He might not have had time anyway, for the police arrived to warn us that if we drew attention to the kayaks, there was more chance somebody might tamper with them. "It's best to keep them hidden from sight as much as possible." We cut short our fun, drained the kayaks, and carried them from the shore. Meanwhile, Bjarny continued to be the object of ribaldry, and celebrity, until Salóme gathered him up and took him home for a hot bath.

Salóme's sister escorted the rest of us around Reykjavík by taxi. Her father pointed out a water tower, shaped and painted like a glass of wine. "It's near a Heimaey refugee area, refugees from when Eldfell erupted," he explained, as the taxi weaved through a housing estate between wooden Norwegian-style houses. "People think Heimaey refugees drink heavily. It's a controversial joke, how it's made to look like a giant glass of wine," he grinned. "Some people don't think it's funny."

The whole family, with friends, came to send us off from Keflavík, at nine-thirty, on a misty evening. It was the biggest send-off so far this trip, but I waved goodbye subdued. As soon as I started to get to know people, it was always time to leave again. I had been so happy here. But my despondency was short-lived. Having left everyone behind us on the beach, to our surprise they all appeared again, waving to us from the next headland. They had hurried ahead to watch us pass.

With the mist, and the night, came a peaceful sea with scarcely a breeze. I missed my chart, vanished from my deck in the last few days. Despite that, instead of hugging the shore, we made two, short, blind crossings in the fog, guessing at the compass bearings we would need to hit land.

No sooner had we lost sight of land than our kayaks slowed, the water thick with jellyfish. They were of the non-stinging type

Keflavík and Reykjavík

with four purple rings, moon jellyfish, *Aurelia aurita*. Although small, there were so many close together I could feel their drag against the hull. At every stroke, my paddle thumped into the gelatinous constellation with a soft thud.

Startling me from my observations, Geoff reversed abruptly with a shout. Beneath him, a lion's mane jellyfish, a good two feet across, quivered, and pulsated. Cream-colored tentacles streamed ten feet behind the mass of fluffy translucent brown frills.

Having expected to easily pass Reykjavík in one push, I could not keep awake. Pausing for a break, I pleaded, "If we keep going, we'll be so knackered we'll sleep till evening and end up paddling at night again. Couldn't we sleep now and paddle by day?" Geoff agreeable, we landed.

At the top of the beach stood fish drying racks, beyond which stretched a golf course, an inappropriate place to camp. In the end, eschewing the effort to flatten a place for the tent on the stony beach, we each chose somewhere to sleep in the open. I picked a strip of sand, between ridges of pebbles, and arranged the kayaks close either side of me as a windbreak. With relief, I noticed how the breeze, just stirring, had already begun to lessen the stink of fish.

Voices woke me at eleven in the morning. I rolled over to find a crowd clustered around me. Men and women alike in bibbed waterproof aprons, as from a fish plant, stood over me talking among themselves. Feeling awkward and shy, I murmured, "Hello," rolled over, covered my head, and pretended to sleep.

I must have really slept, for it did not seem long afterward when heavy drops of rain on my face woke me. I pulled my bivvy bag around my sleeping bag and went back to sleep. Waking periodically, I noted the tide creeping up, and persistent rain. Was I high enough up the beach to avoid the tide, I wondered?

Eventually, I had to get up. In the chilling rain, impatient for coffee, I could not light the Primus stove. Carelessly, I had left my lighter handy, out in the open beside the stove, and the wet flint would not spark. When I discovered my matches were damp too, I gave up and looked for Geoff.

"Let's shelter over there," he suggested, pointing along the rain-drenched beach to a lookout shelter on the low headland. The burned-out shell was drafty, leaky, and less comfortable than he had hoped, yet better than staying out in the open. There, whenever we managed to light the stove, it quickly fizzled out. After three failed attempts, Geoff was clearly as frustrated as I was.

"It needs another clean-out. We had better look for a workshop and fix the stove. We need to shop too. It is already five-thirty. I guess we'll have to spend another night here."

It seemed a long walk into Reykjavík before we located a workshop, beside a yard with wooden fishing vessels propped up in stands. There, a man stood welding the chassis for a camping trailer.

"I plan to drive it all the way around Iceland," he explained.

"A great idea," I approved. "We're doing something similar by kayak, with a tent. A car with a camping trailer is far more practical."

"I work mostly with wood," he gestured at all the boats in the yard, "you know, repair work. I'm building this in my spare time."

With a borrowed wrench, the stove came apart, and we cleaned all the parts, removing masses of black tarry matter. Installing a length of clean wick, we reassembled the stove, our hands grimed and smelling of gasoline. Back at the headland it was time to pitch the tent. As we spread our damp bedding, Geoff commented sardonically, "If we hadn't been too lazy to do this last night, our sleeping bags might still be dry."

When I fetched the stove to boil, I saw something was missing. "Oh no, the key's not there!" I sighed in dismay. The key, needed to control the fuel valve to the burner, usually hung on a

Keflavík and Reykjavík

chain to one of the bars that supported the cooking pot. Whenever we removed the key to fold the stove, the chain kept it from straying. To service the stove, we had taken apart everything, including the pot supports.

"We must have left it at the workshop. Well, it is much too late to go back tonight. We'll have to pick it up in the morning."

Geoff groaned. "So that means yet another late start. What a waste of time and energy. What a stupid stove. We should have thrown it into the sea long ago."

I am sure Geoff blamed me for bringing a defective stove. When I bought it new for this trip, it worked flawlessly. I chose not to point out that it was he who had packed away the stove, leaving the key behind. We may have lost time due to my stove, but he had himself to blame for the extra delay.

"Do you have your pliers handy? We should be able to make it work for tonight." Once alight, the stove partly redeemed itself by burning powerfully and boiling the water quickly.

Next morning, showers of spray awoke me early, the tent leaking under pouring rain. I buried my face and slept on, waking again after the rain had stopped. To my relief, when I lit the stove, it still worked properly. Geoff ran ahead to fetch the control key and joined me waiting outside the store we passed yesterday. Rewarded for our patience, when the shop eventually opened, we bought fresh food. Finally afloat, we turned from the golf course and the malodorous fish racks, racing away with the wind at our back.

Near Miðbakki, Reykjavík's harbor, a magnificent four-masted ship, departing under partial sail, drew our attention. Was it a restored relic or a replica? We idled, watching it pass. There was something majestic about such a large vessel forging so softly past, without all the clattering thrust of a modern ship.

A couple of hours north, toward Akranes, I spotted a whale and noted the coincidence. "Geoff, on the map, this fjord is

Hvalfjörður, *Whale Fjord*." I did not realize how incongruous that was until later. At the head of Hvalfjörður, on the site of a former American naval dock, stood the only commercial whaling station still operating in Iceland.

We did not feel as if we had left Reykjavík behind until we reached Akranes, ten miles on. There we ran into tricky offshore breaks, one of which caught Geoff by surprise and carried him all the way to shore, dumping him onto rocks. The incident reminded us not to get complacent.

Stopping just before the town to fish, Geoff realized the wind was once again too strong, blowing him along so fast his weighted hooks trailed behind at the surface. Landing emptyhanded, we began to set up the tent. At that moment, I discovered I had left behind the A-pieces that held the frame together. It was my mistake for not double-checking in the rush of packing. As annoyed as I was, there was no question of paddling back to get them. I used the spare I carried with me at one end. At the other, I improvised with a length of string and a piece of rubber scavenged from the beach. It did the trick. The tent standing, we stretched our legs and explored the town.

By seven next morning, Geoff had already made cocoa before hurrying into Akranes to buy batteries for his cine camera. I wrote a letter and mailed it, so family and friends would learn how far we had come. I would have to wait until the end of the trip to find out how everyone fared back home. It is interesting how we carry mental images of places and faces with us when we travel. Remembering how they were, and how they looked, we expect them to remain the same, only learning how they have changed when we see them again.

I was extra careful packing everything up here, determined not to leave anything else behind, double-checking the beach before we launched. Leaving Akranes, the wind that continued to push us along carried a strong, fishy stench. Whether that came

Keflavík and Reykjavík

from fish processing or cement manufacture, I could not tell, although fish plants seldom smell so strongly. The cement factory, built in 1958, uses shell-sand from Faxaflói, instead of chalk or limestone. Formed from calcareous seashells, the sand pumped from offshore is more than ninety percent calcium carbonate. Shell, with all that comes ashore with it, must surely smell fishy when roasted. My thoughts straying, I began to paddle automatically.

OUR 1/4 PINT STOVE.

GEOFF COOKING OVER BEACH FIRE.

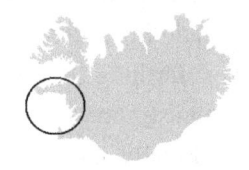

20
Faxaflói

Faxaflói is the bay between Reykjanes, to the south, and Snæfellsnes, fifty-five miles northwest. In January 1942, a German submarine, U-132, torpedoed a treasury class United States Coast Guard cutter in Faxaflói. Twenty-six men died. The vessel capsized the following day. An American destroyer fired on and sank the unsalvageable wreck, and the *Alexander Hamilton, WPG-34*, became the US Coast Guard's first casualty in World War II. The wreck lies twenty-eight miles offshore in this bay.

We had already paddled eastward into Faxaflói to reach Reykjavík. From there, continuing northwest along the back of the bay past Akranes, the peninsula that defines the northern end of the bay was in view in the distance. This peninsula, Snæfellsnes, was sometimes visible even from Reykjavík seventy miles away. The captivating snow-capped volcanic cone of Snæfellsjökull marked its western end.

The nearest coast being low, both behind and ahead, I welcomed the distraction of a small whale that surfaced repeatedly ahead as we sped with the wind across Borgarfjörður. This fjord is the estuary of the Ölfusá river, the biggest river in Iceland by volume. Nowadays marking a county boundary, the river defined the administrative boundary between the south farthing and the west farthing from as early as AD 965, when Iceland divided its administration into quarters.

Finally, we reached a group of grass-topped islands and skerries. The largest islands had small sandy coves with sickle-curved beaches.

"Look, it's yellow sand," I called out surprised, as we skirted the shore. "I love that color, with the green. What a change from black." I recalled the somber mood of the southern beaches.

We landed for a brew at Akrar, the name meaning *fields,* and ate the last of our pudding. "Do you think we should keep going, to the end of these sandbanks, and cross to the next peninsula? That would be, what, another ten or twelve miles? I'm getting chilly here."

"Yeah, but it's going to be wide-open there, and this wind feels stronger than before."

"Well, let's paddle a bit farther, see what it's like. If it looks too much, we can always stop this side."

By the next cove we could see windswept whitecaps on the open water ahead. Discouraged, we turned back, realized the potential of a ruined farm building, and set up the tent in its shelter.

"This isn't so bad! I'll get a brew on." As I took the gallon plastic milk container of gasoline, to refill the stove, I realized how lightweight it had become. "We'll need to get more fuel soon." I held up the bottle and shook it for Geoff to see.

I primed the stove, igniting the gasoline in the circular trough around the pipes. It flared into yellow-blue flames, billowing sooty smoke. I waited for the flames to subside, easing open the key to allow the vaporized fuel through the jet. Only a dim candle-flame of dark yellow appeared, flickering weakly. When I opened the key fully, I saw no change in the flame.

Faxaflói

"Ugh! This is so frustrating. It's playing up again." I took out the *pricker,* the tiny point of wire on a flat metal handle which came with the stove, to clear the blockage in the jet. I scratched the point around the jet hole, struggling to poke the end in. "The jet's blocked," I announced bluntly. Geoff grunted in contempt and climbed into his sleeping bag where he feigned sleep.

I worked at the jet aperture until I had cleared all the black deposit that had almost closed the hole. Then, pouring a little more gasoline into the tray, I lit it and watched the flames envelope the burner. When the fuel had burned to the last, the pipes hot, the flames dwindling, I eased open the key again. I got an even smaller candle-flame than before. So, the blockage in the jet had not been the main problem. I fussed with the stove for about an hour before giving up. I needed the tools to take it apart.

Geoff slept soundly, so I left him and carried the stove with me to the nearest farm. There, I asked an elderly lady outside sawing wood if I could borrow tools.

"Go to the second door of the house down there," she pointed, "beyond the farm buildings."

The man who answered the door said, "Try these." He handed me a choice of heavy-handled wrenches. The brass nut on the stove had become so burred over, it was impossible to get a grip with a wrench.

"If that's no good, have a go with this." He held out a huge pair of pliers. That did the trick. Out slithered the guts of the stove: the blackened intestinal wick, with a gush of dark, dirty fuel.

"Here," he passed me a handful of lamp wick to dry my hands. I replaced the stove wick with fresh and tightened the nut. A group of boys had gathered around to watch.

"How long will you stay in Iceland?" one asked.

"All summer. We are kayaking around. We started at Seyðisfjörður."

That generated excitement. One boy ran off, returning before long to invite me for coffee. In the kitchen he hurriedly unfolded a newspaper onto the table. He pointed to an article with two large photos of Geoff and me, and another with a large photo of us leaving Keflavík. The story about us, which I was unable to read, continued, on the back page.

"Where are you camping? I'll fetch your friend," the man offered. He drove away and soon returned with Geoff, who entered grinning happily. In his arms he carried our wet kayaking clothes.

"They've offered to dry it all!"

Into the commotion, someone carried a book. "Will you sign this for us?" It is a common custom in Iceland to keep a guest book. The handsome hardcover book and a pen were ready for my attention right when the door opened again. An attractive woman bustled in. She glanced quickly around the room and greeted everyone in Icelandic. Switching seamlessly into English, she explained that she was the daughter of the older lady, the one I had met outside when I first arrived.

"I've just driven home from Reykjavík. Tomorrow morning, we expect seventy-five, a hundred people here. My father died, and his funeral will be tomorrow."

Geoff and I walked back to the tent, where I refilled the stove, primed, and lit it, expecting it to work. Instead, I only got a tiny glowing flame.

"This is ridiculous. It must be the fuel." Then, something occurred to me. The sickly-sweet smell in the air was reminiscent of burning plastic.

"Geoff, maybe the gasoline dissolves the plastic bottle, and the burning plastic gums up the works." That would make sense since our problems grew worse when we ran low on fuel. "We need to find another container." Rain blowing in again I set aside the project, abandoning any hope of a hot meal.

Faxaflói

Next morning the farmer woke us. He had brought a bottle of paraffin (kerosene) and insisted I took it for the stove. I explained how we used a different fuel and let him smell the stove. When he came back later, he brought our dried clothes and a rubber hose. He pushed one end of the hose down into the fuel tank of his car and sucked on the other end to get the fuel running. Spitting out a mouthful of gasoline, he stuck the pipe into our empty fuel bottle and siphoned it full.

The clean fuel burned in the stove with no problem, which reinforced my suspicion that our problem lay, not with the stove, but with the fuel, and quite possibly because of our bottle.

Packed up and ready to leave, Geoff looked down at his feet and said joyfully, "For the first time in weeks my shoes feel only *a little* damp!" Our footwear normally stayed soaking wet all day, and impregnated with salt, rarely dried overnight. Our clothing seldom dried fully either. It was a treat to start the day feeling so comfortable.

The weather cleared, almost windless. Along the sea horizon bubbled a line of fluffy cumulus clouds in an otherwise blue sky. We aimed our kayaks at the distant mountains across a slow heartbeat of gentle swell. For once, the calm conditions let us keep up a conversation, paddling side by side.

We could seldom chat like this on the water. I usually focused inward, tuned to the beat of paddle strokes and the rhythm of the waves, taking note of the changes in weather, watching seabirds, and scanning the cliffs. The surroundings often distracted us from talking, even when the sea was calm enough to paddle close together.

Reaching land again near Löngufjörður, we decided to take a break on a small island and paddled behind it to look for a surf-free landing. There, a current carried us swiftly past. "Could it be the tide from Löngufjörður taking us? It's going our way." We landed despite and stood studying the water. "It must be the tide. There's not enough wind to help us that much."

My eyes, scanning down the beach, settled on a strange object. What was it? Shaped like a large mound, it had structure to it. "Look over there. Is that a dead whale, what's left of one?"

We strolled along the yellow shore to verify it was indeed a whale carcass. It was mostly flesh-free, although not yet completely reduced to a skeleton. Pitted and pecked amber chunks of matter dangled here and there from the bones. When prodded with a driftwood stick, everything still held together surprisingly strongly. Just as when pushing against the side of a car, the whole thing budged a little when shoved. It settled right back again afterwards, gassing off an even more putrid stench. Flies buzzed noisily and settled.

"I'd like to take a whale bone home with me," I mused. "Nothing huge. Just a small vertebra would do."

"I'd like one of those big ones we saw, to use as a seat. Can you imagine having a bunch of those huge bones in your garden as chairs? That would look so cool."

"You'd have to find old ones, otherwise they're much too oily. They're disgusting like this." I jabbed again with my stick, making the rib cage shudder. A waft of rancid odor embraced me amid a cloud of flies. "Ugh! Let's get something to eat."

By the time we relaunched, the tide raced past with breakers everywhere, the current carrying us until we reached a spit. Seeing clean regular swells breaking along the spit we glanced at each other, questioning. The temptation was too great; we stopped to surf.

The rich color of the sand glowed through the clear water. My kayak lurched over each glassy slope that passed under me as I waited for the right wave. Exhilarated, I surged forward on a swell as it steepened, turned, and rocketed diagonally across the wave toward the end of the spit, trying to reach deeper water before the

Faxaflói

break caught the kayak. My kayak, despite its cargo, flew along with each ride.

At play, we had no reason to stop until an extra-hefty break tripped and pounded Geoff. He rolled up and paddled back out smiling, saying, "Refreshing! You should try it!" I laughed. Rolling was the last thing I wanted to do. But I saw past Geoff's stoicism. He looked cold, his clothes saturated. We should leave. On the move, he would warm up once the water drained.

The mountains along the peninsula seemed to grow and sharpen, gradually unveiling themselves as the clouds broke and dissolved. More peaks appeared as the sun sank lower. We noticed the moment the sun slipped behind the range, casting us in shadow, the air instantly cooler. With just one last valley behind us in sunlight, I began to look for somewhere to land.

"Look! There's someone waving at us, over there," I pointed. Two figures stood on the shore, in a cove beneath a lighthouse. "Let's go see what they want."

"Why don't you go on ahead? I'd like to try to catch a fish before I land."

"Okay," I agreed. As I left him, I quipped, "Make sure you stay outside the two-hundred-mile limit!"

Thor and Stella, a couple a little older than me, greeted me ashore. "We are part-owners of that house under the lighthouse," Thor explained. "We drove out here this evening for a break."

"And then we spotted you and tried to wave you in," continued Stella, "I am so glad you saw us. You can stay with us of course. I'm sorry now that we didn't bring more food."

Thor said he was an engineer, and that he often visited New York, and London. As we all talked, I saw how close he and Stella were. They made a good couple.

Later, we all sat together talking, sipping Scotch, and watching the old Snæfellsjökull volcano progressively shed its cloud to reveal a cone-shaped peak capped with ice.

"It looks like an old elephant, when we see it from our home in Reykjavík," Stella said.

"It looks like a headless camel from here," Thor grinned.

Stella continued, "I think all the mountains along here look best from a distance. They're just like people, always more perfect when you don't look too closely." I hoped our bedraggled appearance had not inspired her comment. I felt scruffy.

Later, I slept on a bed by a window overlooking the sea. When I awoke, thinking I had only just closed my eyes, I had slept soundly for hours. Stella and Thor were about to head out for a drive. "Stay!" Stella ordered. "Treat the place as yours. It will be a break for you. Any food of ours, consider it your own." The house seemed empty when they left.

I peered into the bathroom mirror. Recalling Stella's comment last night, to not look too closely at people, I thought to tidy my appearance. I washed my hair, trimmed my beard and fingernails, and felt better. When I hand-washed my clothes, I was disgusted by how dirty they left the water; surprised by how bright their colors looked afterwards. Meanwhile, Geoff cooked.

Outside, the sea looked flat, and dully matt, beneath sheets of rain. The rattle of the downpour made the house feel extra cozy. "There's little point in leaving before our clothes are dry is there?" I asked, hopefully.

"Not really. We could have caught the tide around the end of the peninsula but it's too late now. Let's use the opportunity to get stuff clean."

Thor and Stella bustled in from their Sunday drive, only to pack ready to leave again. "Please, empty the fridge when you go. Stuff won't keep. And have a great trip!"

When their Range Rover left this time, it was for Reykjavík with their distant view of the old elephant.

"This is the life," sighed Geoff, breaking the silence as he turned, a cold lager in each hand, one for me. "Here's to Thor and

Faxaflói

Stella! And to a successful second half of the trip too." We clacked our glasses together.

Geoff turned on the radio before investigating the fridge again. His muffled voice sounded surprised, "Did you notice they left a hunk of beef in here? And tomatoes, cucumber, cheese, bread. Oh, and sausages. And there's milk and coffee." He was happy. As I stood, relaxed and clean, the radio murmuring in the background, staring out at the rain through the window, Geoff added, "The only thing missing is a good woman!" That reminded me how far we still had to paddle before we could see our girlfriends again.

Later, when it was obvious the rain would persist, I braced myself and said, "Okay, we should get going. It's calm. It's settled. There's not much wind."

"Oh, come on," Geoff protested, "there's little point leaving when it's pelting down like this. It's comfortable here. Enjoy it. The weather might get better tomorrow."

"Or worse."

"Really?" He raised his eyebrows in disdain.

"Okay then," I conceded. "Another night here, but we should leave early tomorrow if we can."

A comfortable bed in the dry was admittedly a luxurious contrast to a night in a rain-lashed tent. In the early morning, we locked the house and left through sheets of rain. The jökull elephant, or headless camel, lurked beneath deep cloud. Little by little, as we paddled, we saw peaks materialize one by one; bit by bit as if aggregating from coagulating raindrops.

"This is spectacular," I enthused, as still more mountains shed their veils. The uncovering added a dramatic vertical dimension the early morning view had lacked, under its low and glowering ceiling.

Down at sea level, seals adorned the ledges, arching like grubs as they strained to see us better. The lava fields of Búðahraun, and the solidified lava flows from Snæfellsjökull, formed cliffs along the shore. The embossed overhangs and huge gouged hollows, arches, and columns tortured my imagination. I could see ghastly forms and faces glowering at me from these cliffs.

Looking up, I could barely distinguish between the ice cap and the brilliant cloud. Yet somewhere above us lay the upward route described by Jules Verne, as first seen by the exploring party in his story, *Journey to the Center of the Earth*.

From whence we started it was impossible to make out the real outlines of the peak against the grey field of sky. All we could distinguish was a vast dome of white, which fell downwards from the head of the giant.

Following in the footsteps of the fictional explorer, *Arne Saknussemm*, this intrepid party scaled the volcano before descending into the crater. From there, dropping down a shaft, they explored an underworld labyrinth of ancient seas populated by prehistoric creatures.

After amazing adventures, they escaped from the bowels of the earth, their raft lifted on rising magma. A volcanic eruption tossed them onto the sundrenched slopes of Mount Stromboli, on the Aeolian Island, Stromboli, north of Sicily, far from where they started here on Snæfellsjökull.

I had no desire to follow their route. Memories of slithering through caves would suffice. Crawling through wormholes to the center of the earth could wait.

We crossed Breiðavík as the tide slackened, to land behind an almost hidden harbor arm at Hellnar. In its shelter, four small fishing boats lay grounded on the sand against the wall. Geoff, in need, had rushed ahead to land, quickly taking shelter behind one of those boats. I followed more carefully, wading through the deep layer of pungent seaweed.

Faxaflói

Someone saw us arrive. He drove his car to the harbor to greet us, and to invite us for coffee. I looked at his car. "We can walk, we're soaking wet."

"Oh, that's no problem," Thorvadine dismissed, gesturing for us to get in. Still, I cringed when seawater from my fleece pooled on the seat and around my feet.

Thorvadine drove us from the bay to his farmhouse, less than a half-mile up the hill. We could have so easily walked.

"My father and I fish and farm. We fish in spring and summer and farm all year-round."

"Do you use nets?" asked Geoff.

"Hand lines. Today's no good. The tide would be too strong. The lines wouldn't reach the bottom."

"I have that problem when it's windy," said Geoff, perking up. "I hadn't thought about the tide being a problem too."

"It probably isn't if you don't have an anchor," I chipped in.

"There's good fishing near Malarrif. That's on the other side of two rocks that stick up from the cliff," Thorvadine suggested. "You'll see on the map how the volcano forms Snæfellsnes, making the end of the peninsula almost circular. Malarrif is at the southernmost point of that rounded end."

We entered the kitchen. "Oh, here's my mother," Thorvadine said introducing us. I could not help wondering if all Icelandic women are born to be beautiful. Part of the attraction was their self-confident demeanor: a readiness to make eye contact, to break into a smile, and to maintain both.

She began to set the table with bread, cheese, and tomato, and invited us to sit. Then, she carried steaming mugs of fresh coffee to the table, setting them down in front of us. "Can you show us where you have been?" she asked, leaning over, and opening a map of Iceland.

"Of course," I began, "we arrived here, on the ferry to Seyðisfjörður," I pointed. Over the weeks I had repeated the sequence so many times, and had my pronunciation corrected so many times, that I could roll out most place names quite confidently. That is not to say there were no pitfalls. I had discovered that depending on who I talked to, and where they were from, there were subtle differences in how people pronounced even a simple place name such as Vík: *Vick, Veek, Wick,* or *Week.*

Thorvadine drove us back down the hill to the harbor to see us off. Refreshed and energized we paddled briskly around the cliffs to Malarrif where three fishing vessels from Ólafsvík were trawling. When one of them altered course to intercept us, we stopped. With the boat lurching and rolling in the swell, we kept backing away for safety. The crew peered down at us from the deck, at least ten feet above us. "Would you like coffee?" one called out.

"Yes please," I shouted up, "I would love a cup, but how?" My question hung there. The man had vanished without reply.

"Here!" came a shout. I looked up again. He was back, holding out a mug. Is he serious, I wondered? I maneuvered my kayak as close to the hull as I dared and turned broadside. My paddle almost caught me off balance at each downward roll of the trawler. As it pushed outward it left no room for me to place my blade.

"Hey! Look here, look out!" The shouts came from above. He was reaching down over the side. As the trawler paused at the low point of its roll, I grabbed the hot mug in my free hand. Moments later the hull was rising again beside me, as rapidly as if I were falling in a lift shaft. I almost lost balance. Fumbling, I could not brace against the water with the boat so close to my other side. Wobbling wildly, I hooked my paddle behind my shoulder and one-handed pulled the kayak sideways to get more space. I was trembling, barely balancing.

"Good job!" complimented Geoff, who had declined the offer of coffee and sat at a respectable distance to watch the action.

Faxaflói

I sipped my coffee while Geoff called up questions, and the crew stood along the gunwale grinning down at us. They were fishing for plaice. Once I had finished drinking, I waited till the boat rolled sideways toward me again, before flinging the empty mug upward to a perfect catch.

Farther on, with Snæfellsjökull almost clear of its cloud cap, another fishing boat spotted us and powered across. This crew also offered us coffee. After successfully repeating the same tricky procedure, I wished I had designed a less tippy kayak.

"What are you fishing for?" I asked, mug in hand.

"Cod," came the reply.

After all this coffee, I needed to pee, so we landed on a ledge of lava. "Do you think those two boats radioed each other to say we were coming?" I asked Geoff.

"Yeah, I expect so. If the second one heard about the coffee from the first, I imagine they hoped to see you fall in!" I grimaced. I had barely balanced. Geoff continued, "And they never bear any grudge about the cod war, do they? It was friendly of them to come over."

Once we had paddled past the two lights, and rounded the corner to Skarðsvík, the northwest peninsula came clearly into view. At the northern end of Breiðafjörður, those huge cliffs across the bay were at least thirty-five miles away.

"They look so close, don't they?" I mused. "How about paddling straight across instead of going around the bay? That would save time." I could tell Geoff was weighing up the idea.

"In any case, if we paddle a bit farther today, the crossing will be shorter, that is, if we decide to go for it." The coast continued northeast for another five or six miles before the ideal point.

We began to pass the village of Hellisandur, before spotting the transit stones of an old landing at historic Keflavík. "Let's pull in there." The leading marks steered us to a rocky, nevertheless sheltered landing.

Iceland by Kayak

On our arrival, when all I wanted was a few minutes of peace to change into dry clothes and set up camp, children came running from all directions to engulf us. Once again, I became perturbed, my every move followed by attentive eyes. But in time, the children, sufficiently bored, or their curiosity satisfied, drifted away like loose leaves in the wind.

We were warm and dry before a young couple came to ask what we were doing. As we talked, a little girl approached hesitantly and shyly handed us a grocery bag with a note. In the bag was food, and a thermos of coffee. The note read: *Please return my thermos. I live in the red house.*

The generosity, and hospitality, warmed me yet again. After we had eaten, and finished the coffee, we returned the flask to the red house where the little girl's mother, Sigríður, invited us in for more coffee. She spoke quietly, in good English, translating her husband's conversation as he cooked us bacon and eggs.

"He works on fishing boats, often as a cook, sometimes on the lines."

"Hand lines?" asked Geoff for clarification.

"Well, they use electric reels now, so they're hauled up automatically."

When at last we stood to leave, Sigríður looked at me and insisted, "You must come over for lunch tomorrow." It was a lovely invitation that I looked forward to.

By the time we got back, it seemed almost dark inside our tent. "The nights are not as light as when we started," I noted, thoughtfully. "We need to get a move on. We should be halfway around, but we're not yet, are we?"

I looked at the calendar to see how long we had taken so far. "We started on the twelfth of June, and it's already the eighteenth of July. We've only a month left, no matter what. If we cross straight over Breiðafjörður from here we could save a couple of days, or even a week depending on the weather. But then again,

Faxaflói

Breiðafjörður is supposed to be special, full of little islands with strong currents and loads of birds. It would be a pity to miss it."

I marched my dividers across the map to measure the distance around Breiðafjörður, and across it. "It's about one hundred to one-ten miles around, depending on our route. Cutting straight across would be only thirty, or thirty-five miles, and we'd finish that all in one go." We would have to make that decision here.

CHILDREN PLAYING BY THE KAYAKS, HELLISANDUR.

Iceland by Kayak

MAP 7. NORTHWEST ICELAND.

21

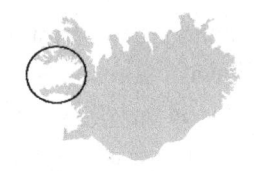

Breiðafjörður and Beyond

The nineteenth of July was hot and sunny with no appreciable wind. The Snæfells-monster, the icy-topped mountain, was in full spectacular view eight miles inland. While Geoff looked for a bank to cash a traveler's check, and shopped for food, I asked around for a weather forecast. What I heard was inconclusive: "Sunny, wind from the northeast force two to four, Beaufort, but who knows, the weather's never that predictable here."

Undoubtably, the pleasant weather made me feel lazy. I looked forward to seeing Sigríður during her lunch break. Content to relax in the sun waiting, I lay there, breathing in the fragrance of the warm grasses, flicking away the occasional summer flies. Geoff, on the other hand, seemed impatient to leave while the weather held.

At lunchtime, Sigríður greeted us enthusiastically, saying, "At the fish factory this morning, "all the other girls were saying, 'Oh, so they are staying at *your* house?'"

She said that she was planning to visit London with a friend in mid-November. Would we be able to meet them there?

Iceland by Kayak

When Geoff and I launched with a new cargo of food, including two fresh fish from Sigríður, it was onto a silky sea. Our kayaks slid smoothly and effortlessly, beneath an endless blue sky, in dazzling sunlight. For the first time since leaving the east coast, it was warm enough to be comfortable without jackets.

After a few hours we noticed a group of fast-moving creatures approaching, fins surfacing in close groups, all traveling together. I could not tell how distant the creatures were, or their size until they grew closer. When, occasionally, one leaped high out of the water, twisting in the air to expose its pale underside, I recognized they were dolphins, some faster and a little larger. We watched for a long time as they raced toward us from our right, while still more converged from our left.

A small group seemed bold, surfacing ever closer to our kayaks, slowing to match our speed. After a while they began to surface within yards to our left, and began broaching silently between our kayaks, holding their breath. Stunned, I was delighted by how close they came, for our kayaks were never more than twenty feet apart.

When the dolphins cruised away, I focused again on the cliffs ahead. They already seemed quite close. "It looks as if we only have about ten more miles left," I judged, from the clarity of the cliffs. Yet the cliffs soared to more than two thousand feet. With today's clear visibility, the proximity was deceptive. We still had another twenty-five miles to go.

I recalled the first time I paddled across the English Channel, feeling perturbed when the low French coast never seemed to get any closer. Having learned what to expect; imperceptible progress, I could be confident that we were in fact creeping toward our target.

The familiar tune of *Sunshine Superman* began to measure my paddle strokes once again. I did not know why it had stuck in my head for more than a month, but its metronomic pace matched

Breiðafjörður and beyond.

perfectly today. Having seen Donovan perform at a hall in Bristol I recalled how, under his mop of curly hair, he had sat cross-legged on the floor with his acoustic guitar on his lap. Relaxed, he had chatted conversationally with the crowd around him, as if in a small room with a couple of friends.

With satisfaction we drew close to shore at the northwest peninsula. This region, otherwise known as the *Westfjords*, spreads like a hand from the main body of Iceland, joined by a wrist just four-and-a-half miles across. If not for that narrow connection, it would be its own island.

It had taken us eleven hours in all to cross Breiðafjörður to Rauðasandur, and we spent another hour skirting the sandy shore before finding a better place to stop. We ran ashore onto pebbles, where rocky ledges broke through the sand and a braided stream ran down the beach.

After we had parked the kayaks up the beach I laughed gleefully. "Geoff! For the first time, I'm completely dry getting off the water. Drier than when I set out! I love it." I also felt smug about completing in one step what would have taken us much longer, had we paddled around the perimeter of Breiðafjörður. Skipping joyfully from boulder to boulder across the stream, toward a good spot for the tent, I slipped and landed on my back in the water, unhurt but drenched. "Oh no! I cannot believe it," I groaned. "I was so dry!"

"At least it's fresh water. It'll rinse the salt out." Geoff's voice sounded thoughtful, not laughing. I looked at him, wondering. He sat nursing his wrist.

"What's up?" I asked.

"A little twinge," he replied. "I must have tweaked it coming across today. I expect my wrist got too cold when I left my jacket off."

"Better go easy for a couple of days," I suggested. "Wait and see how it feels." We could not afford for it to get worse.

Iceland by Kayak

Having cooked before pitching the tent on the rocks, I found it impossible to sleep in the tent. In the sun, it was too hot. Yet outside, seaweed flies tormented me. Unable to relax, I spread all my stuff out to dry, and inspected my anorak.

The seam under the arm had given way, leaving a hole. The tough seam thread had pulled apart the stretchy fabric, which had become too weak and stressed to hold a stitched repair. The other underarm looked the same. I picked up Geoff's anorak and saw it was coming apart under the armpits too. The shoes I wore for paddling also showed signs of abuse. The soles were separating from the canvas uppers. They did not fare well, constantly wet. I had already replaced the laces, long since rotted, with wiry strands pulled from polypropylene rope scavenged from the beach.

I sat back against a rock to study the view. Fifty miles away across the bay, Snæfellsjökull looked spectacular, bright with snow. A cloud hung over it, barely touching its summit. It was such a serene scene; Geoff took out his cine camera to film. But a cloud over the summit? That was a warning sign. The Admiralty Pilot suggests that a view of the summit standing clear, unclouded, signals fair weather. It warns to expect a change when clouds start to form on it, usually a gale. Should we prepare for windy weather?

We began planning the next leg of our journey, around the southwestern corner of the Westfjords. At the western extremity, the tide would run at up to four knots causing the tide race, *Látraröst*, where the north-going stream joins water circling west from Breiðafjörður. After that, our course would trend northeastward for the next eighty miles or so. I was keen to get going.

"If your wrist isn't too bad, can we paddle around the corner; ten miles or so? Potentially we could go as far as Kollsvík on this finger of land, what, twenty miles, before the next fjord. But we

Breiðafjörður and beyond.

could stop at any one of the bays before that. What do you think? Are you up for it? We should get help from the tide."

Geoff was amenable. We launched in sunshine, onto calm water, but as we began paddling, I sensed the tide against us. That puzzled me for I thought it should have been in our favor. Was there an eddy here? Briskly passing Keflavík with its refuge hut, I visually checked the beach which looked good for landing. "Next time," I promised myself, hoping I might come back one day.

From there on, the steep Látrabjarg cliffs climbed to fifteen hundred feet, banded red and grey, with creamy white bird lime oozing from ledges. Millions of birds surrounded us. Huge rafts of guillemots and razorbills coated the sea, clucking, cawing, and coughing. As our kayaks ploughed a path, they scattered in all directions, scurrying aside, or diving into the clear water. I could see their dark bodies flying gracefully under me, tiny bubbles trailing behind them. The surface swirled with debris; rancid, with oily slicks of bird lime congealed with waterlogged feathers and down.

The guillemots had chicks with them on the water. The adults made a gargling sound when we drew close, calling the chicks to them. I tried to imitate that sound and encouraged two or three chicks to follow me instead. The parent bird took offense and flew after me, splashing across the surface, protesting loudly. Curious, I tried again, and again, trying to perfect my call as we cut our path through the endless rafts of birds. It always offended the parent birds. Proud that my false call sounded good enough to deceive the chicks, I empathized with the parents' distress and gave it a rest.

We began to feel the pull of the current. It accelerated us into a tide race which rushed us bouncing and bounding past rock ledges with basking seals. At the squat, grey, Bjargtangar lighthouse, which stood on the clifftop, we rounded the southwestern tip of land. Not far beyond, we reached the start of the yellow sand beaches of Hvallátur, which name I translated, probably incorrectly, as *whale laughter*.

"How's your wrist doing, Geoff? Do you want to land?"

"No, I'm fine, thanks. If you are okay with it, let's just stop for a minute on the water, for a snack, and go for Kollsvík."

The tall cliffs, under which we chose to pause, glowed all shades of green where moss and slime coated the rock. The underlying rock appeared brownish grey with red bands. Seabirds occupied every tiny ledge, kittiwakes clinging to the sheer faces. We bobbed up and down, sheltered beneath the cliff, watching the activity.

At Kollsvík, rock ledges reached out to sea like breakwaters from a curving yellow beach more than two miles long. A low broad valley stretched inland. We ran our kayaks ashore onto a ramp of broken seaweed, where a stream rattled noisily between sand and rock. Above the beach stood the ruins of animal houses. Each stone base formed a crater rim, confining a tangle of wooden beams, corrugated metal, and grass from its collapsed roof.

"These should give us a good bit of shelter," I grinned, looking to Geoff for approval. We set up camp there and began to prepare the plaice Sigríður had given us. We had not time to begin cooking it before a young man appeared beside us. "Hilmar," he introduced himself, reaching out his hand in greeting. "I'm from the farm up there." He was chatty.

"Our family, we're the only farm left here," he explained. "There used to be two others, but life got too hard for them. They couldn't make it, just farming. They needed to fish too, but they didn't do so good." Curious, we encouraged him.

"Eighty people once lived in this valley; we are the only ones who stayed. We keep ten cows; that's the minimum we need. We also have sheep. We let the animals graze where the other farms were too. It's better grass."

"How do the animals fare in the winter?" Geoff asked.

Breiðafjörður and beyond.

"We keep them inside, mostly, unless the weather looks good. Then, we let them out to feed on the seaweed thrown up by storms. It makes good fodder. Otherwise, we feed them hay. We cut the grass, twice in a good year. It's thin, so we spray fertilizer on the best fields." Around me the ground was dry with sandy yellow soil, supporting a very sparse growth that looked unsuitable for hay.

"You know there are shipwrecks along here?" he asked. "One of them was a British trawler, the *Dhoon*. Rescuers tried to winch her crew to safety up the cliffs. I expect you heard about it?" Neither of us had, so he straightened himself up and elaborated.

"There was a winter storm around Christmas 1947, when the *Dhoon* ran onto the rocks, just under the cliffs at Látrabjarg. They sent a distress call saying there were three men lost already, including the skipper, with a dozen still on board. When the farmers heard the alarm, relayed from Reykjavík, they got together and walked out to the cliff.

"There was a place where they could get down using ropes, so some people stayed on the clifftop keeping watch, while others went down and hiked along below the cliff to get closer. Everywhere was slippery where the storm spray had frozen, but from down there they could see people alive on the wreck.

"They fired rockets to get a line to the ship, rigged up a breeches buoy and brought everyone to shore. They still had to get the survivors up the cliff, and they couldn't do that until the next day, waiting for the tide and for daylight.

"The villagers set up a tent on the cliff, for shelter, and to feed everyone and warm them up, before they carried the survivors away on horseback." He stopped and grimaced, shaking his head as if imagining their pain.

"The strangest thing happened the next year. A film crew came out to make a documentary about the rescue, and when they went up onto the cliffs to film, they spotted another trawler in trouble, the *Sargon*. That wrecked too, so they filmed a real rescue. I think they got six men off alive that time, from the *Sargon*.

"It's a famous documentary, at least around here," he added. "Yes, It's a hazardous coast for sure. Shipwrecks. Drownings. That's why some of the fishing companies got together to build shelters, to help survivors."

"We've seen some," said Geoff. "We stayed in one or two on the south coast."

"Yes, the south coast is dangerous too, for sure," Hilmar acknowledged. "In storms, so much spray fills the air, the boats can't see when they get too close to shore. Everything is too low to show up on radar."

After Hilmar left, Geoff began cooking the fish, accidentally spilling it onto the sand. It was late before we ate the delicate plaice, sitting on a log, spitting out the grit and fish bones, watching the sun go down.

"Right, let's start early tomorrow, catch the morning tide and take advantage of this weather." That agreed, we settled for the night. Tired, I fell asleep at once, and overslept. In the morning, Hilmar appeared with a guest book in his hand drily pointing out the obvious: "I see you haven't gone yet." He passed us a pen to write in the book and seemed pleased with our entries. "My father," he invited, "would like you to join us, to eat." Delighted to accept, we crossed the fields with him to meet his parents and grandmother at their house.

They treated us with fish, with home-grown potatoes, followed by rhubarb soup and cream. Everything was from the farm, except for the fish.

"We keep a boat in Patreksfjörður for fishing; the fjord to the north," Hilmar explained, picking up a coffee-table book and passing it to me. "This is about Iceland." It showed photographs of places we had seen, including Höfn, Vík, Vestmannaeyjar, and

Breiðafjörður and beyond.

Reykjavík. I pointed out those I recognized, taking pride in having visited each. Finding Akranes, I asked about the tall chimney. Hilmar confirmed that the strong smell came from the cement factory burning shells, not from the fish factory. "We make all our own cement in Iceland."

Realizing, if we talked any longer, it would be time to eat again, Geoff stood. "Well, we really should get going. Thank you for everything!"

Everyone started to move. "Please, take this with you!" Hilmar's mother handed us three bottles of milk. "We have plenty. And here, these are fossils from the north coast. Take one each. There are only a handful of places in Iceland where you can find fossils. The main geology is fire rocks." I took one and looked closely at the grey rock with its embedded gastropod shell. The rock looked like limestone with clay in it, or lime-rich mudstone. Growing up, I used to search for fossils. Thanking her, I wondered if we would see anywhere with that same kind of rock.

As we cut back across the fields, I reflected on how lonely it must get here in the winter, snowed into the valley with little daylight and only family members for company.

Launching, we paddled in sunshine to the mouth of Patreksfjörður where we greeted a man standing in his small open boat; lining for cod using a reel mounted on the gunnel. He explained how the winch worked, and how, in the past, fishermen hauled the line hand over hand to bring in the cod. Cod four feet long could weigh ninety pounds. That made handlining a tough job. "This makes it much less work!" he admitted. "That, and the cod nowadays are smaller. Here, have these," he handed us two fish. "Take more," he encouraged.

"Thanks, but no. Two is plenty, thank you!"

As we pulled away, waving, we watched him haul his line again, his apron splattered with fish blood. Soon, another open wooden fishing boat approached us to ask where we were going.

Iceland by Kayak

This man said he would have given us a tow; had we been going into the fjord. He too offered us fish, which we declined, before he motored on.

"Everyone is so generous," I said yet again. "By the way, did you notice that yacht sailing, over there?" So far, we had seen few pleasure boats. The white sail looked tiny beneath the huge cliffs. We watched over our shoulders as it made its way out, and we just as purposefully crossed the fjord. We only lost sight of it in the mist just before we reached Kópaflaga, on the northern side, where the cliffs rose sharply to fifteen hundred feet.

The view in front, of bold overlapping headlands, leveled to a plateau-like skyline of mountains at fifteen hundred to two thousand feet. During the Ice Age, the ice sheet had shaved the tops from the mountains, while glaciers gouged out the deep fjords. Once the glaciers began to melt, they deposited substantial piles of debris as terminal moraines within the fjords. These account for the low, stony, flat land that juts out from either side of each fjord, sometimes bridging right across. Low and level, offering an easy sheltered landing, these moraines made ideal places for settlements. Every town in the northwest peninsula with the suffix *-eyri* stood on one of these moraines.

Crossing Arnafjörður, we reached Dyrafjörður. On a moraine farther down this fjord lies the village, Þingeyri. "Þing," *parliament*, indicates its role as an ancient legislative meeting place, just as at Þingvellir, *parliament fields,* near Reykjavík. The national parliament of Iceland, the Alþing, began at Þingvellir in the year 930, although two other locations already held regional assemblies. The Alþing is the oldest surviving national parliament in the world, holding its assemblies at Þingvellir until 1798. Medieval remains at Þingeyri, on the other hand, reveal its role as a regional meeting place.

Beyond Dyrafjörður we hugged the coast. High misty cliffs of dark basalt unfolded ahead and folded away behind us. The sun

Breiðafjörður and beyond.

seared the horizon on its low languid trajectory, and began to bulge on the ocean meniscus, until it spread blood-red. Incrementally, the colors dimmed into glowing embers, glowering like smoldering peat, smoking across the horizon.

Behind us too, the sky gradually changed color, dimming, fading from a blaze of brilliant reds and golds into deeper shades of mauves and purples. The cliffs behind us, starved of sun, began to darken to an intense blue-black. I squirmed in my seat to look at the constantly changing palette, which never left us.

As we forged on, the positions of the colors moved. A gash of yellow glowed ever brighter as the sky ahead grew greener. The sunrise developed as languidly and as perfectly as the sunset, the last of the one bleeding seamlessly into the start of the other.

Before the sun showed itself, a line of clouds behind us ignited and began to blush, effusing dark rose pink. Such a strong band of cloud looked foreboding, a fierce gathering wall of weather chasing us. Gaining on us fast, this cloud blossomed taller, wider, and ever more powerfully as it came.

The rising sun dazzled, until the mountains bordering the fjord blocked it. Entering shadow, we rounded the corner into narrow Sugandafjörður, finally finding ourselves against a current. The north-going stream had helped us throughout the rising tide and for longer. The Gulf Stream, which always flows north here, weakens the effect of the south-going tide. It combines with the north-going tide, strengthening it and extending its duration.

Within the half-hour, the pursuing cloud completely swallowed the sun and the sky gloomed to overcast. To either side of us the cliffs, more than two thousand feet high and just a mile apart, stood closed by the ceiling of cloud, boxing us into a dark imposing tunnel. It looked like a cave of doom. Ahead, on the south side of the fjord, past a small fishing harbor a cluster of painted houses stood on a low spit of moraine just above sea level.

Overshadowed, dwarfed beneath the craggy slopes, this was the village Suðureyri.

"What do you think?" Geoff queried as we approached the houses. "How about, just to this side, over there?" A gravel road ran above a steep lumpy beach. I drifted the last yards. It had been a long paddle and a magical night. It was four-thirty in the morning. It did not take us long to lift the kayaks to the top of the beach, set the tent, and begin cooking cod with rice.

AUTHOR, PREPARING TO SEAL LAUNCH.

22

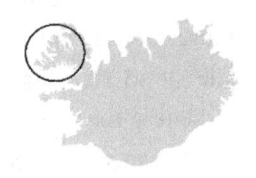

Suðureyri

There were children chanting, the sound so close it startled me awake. Rolling over abruptly, I jarred my elbow painfully against the stones under me. What was going on? There were eyes watching me under the side of the tent, little fingers holding up the edge. The fabric snapped back down as a fit of giggling erupted.

Geoff opened his eyes and frowned. I shuffled in my sleeping bag to unzip the tent enough to see out. A crowd of small children was climbing all over the kayaks. Other children watched to see how I would react. Tired enough to do nothing, I zipped the tent shut and flopped back down, but it was no use. Sleep would be impossible. I got up, and Geoff quietly followed. It was time to escape, to go shopping for supplies.

A breeze was springing up along the shore, wavelets clattering as we returned from the co-op with food. A group of men stood looking at the kayaks and one suggested we should move our tent.

"We expect a very strong wind," he explained, calmly, matter of fact. I shrugged my shoulders. Undeterred by my indifference, he offered, "Come for coffee, we can hear the weather forecast."

Iceland by Kayak

In Guðmundur's kitchen we gathered at the window. His house, beside the water, faced straight across the fjord. The far side stood startlingly close, the imposing wall of mountainside darkening the water.

A lean man, in his late twenties, Guðmundur looked trendy in a white polo shirt under a short jean jacket, with flared denim jeans, and sneakers. As he prepared coffee he explained, "My wife and son are in Reykjavík on holiday."

We all fell silent when the weather forecast came over the radio. Guðmundur focused, his face serious as he listened. "The winds will be strong, from the northeast, not good for kayaks. You should move your tent and your kayaks too. You are welcome to bring them here if you like. It will be more sheltered beside the house." We took him up on his offer and set up our tent where he recommended, right by his house, with the kayaks alongside. He admired the kayaks, before announcing proudly, "I built myself a kayak."

"Really? Can we see it?" His kayak was canvas spread over a broad wooden frame. It looked stable, but his paddle, long with small blades, seemed far too flexible.

"I was inspired by Greenland kayaks, so I built one." Guðmundur stood, his hands in his pockets, looking down at his creation. "I didn't know anything; I just did what seemed right."

We spent the day chatting with him until Óskar, a neighbor, stopped by to invite us over to eat with him and his wife. "That is okay, you should go," Guðmundur encouraged. "They are good people." Barely minutes' walking later, Óskar's wife, Guðrún, *"Gunnur,"* welcomed us into their home.

As we moved through the house, Óskar, a carpenter, stopped to point out through the window. "I've been building that house opposite in my spare time. I was unlucky last winter. I finished the walls and put all the beams in place, ready for the roof. Then, extreme winds, hurricane force." He frowned.

Suðureyri

"I was just leaving the house," he nodded toward his front door, "when my little boy asked me to play with him. So, I stopped, came back, and we played for five minutes. That's when my new house blew down. I should have been in it." He looked down at his son busy on the floor. He repeated, "If not for him, I would have been in it. There was only one unbroken windowpane on the ground, out of all of them.

"So, I was lucky. Unlucky too; I had to start over again." He sighed deeply in resignation.

We enjoyed the whole evening with Óskar and Gunnur, talking and drinking with them until early morning. On leaving, late as it was, we ran into Guðmundur and Joey returning from a dance. We four had stayed up late. But when I saw someone painting the outside of a house, I had to wonder whether they were working late or getting an early start.

Although the weather forecast for the next day also warned of strong wind, it did not seem that bad, so we decided to leave. Guðmundur persuaded us to delay. "I'll drive you up to the airstrip. You'll be able to see better from up there."

We jumped into his Ford Bronco and roared up the steep graded road to the leveled gravel runway, which ended abruptly at the sea-cliff edge. "It's not the best runway," he explained. "It should be longer. Sometimes the planes cannot get up enough speed to take off, so they drop over the cliff at the end and accelerate as they fall. It really scares me. Every time it happens, I think we're going to crash."

He raised a large pair of binoculars to his eyes, scanned into the distance, and passed them over with a gruff, "Yes."

I lifted the heavy glasses and tweaked the focus. The sea beyond the fjord was a frenzy of white. Guðmundur explained, "It's often quiet inside here because the fjord is so narrow. The winds go past, or over. It's completely sheltered from some wind directions, so just because it's calm in town doesn't mean it's calm outside."

Back in town we met the police officer, Sigi, who invited us for coffee with his wife and children. "Me and my wife, and Óskar and Gunnur, also Guðmundur and his wife Parentis, we are all close. Close like a sandwich," he explained.

We had not met Parentis yet, but she was coming home from Reykjavík that evening. Guðmundur invited us to ride with him to Ísafjörður to meet her flight.

The road ran low along the water's edge to the end of the fjord before climbing steeply. The scenery changed abruptly from green grassy slopes to frost shattered rocks and snow. It was like changing seasons within minutes. We usually saw everything from the seat of a kayak at sea level, so I appreciated the contrast, looking steeply back down this narrow fjord from so high up.

The road peaked, crossed the watershed and dropped toward Ísafjörður, where buildings clustered on a hook of low, flat land on the west side of the fjord, Skutulsfjörður. The latter is a small fjord that cuts southwest from the longer northwest-southeast running Ísafjarðardjúp. At its southeast end, Ísafjarðardjúp doubles back toward the southwest as a tail-like fjord named Ísafjörður, more than twenty miles from its namesake town.

Ísafjörður is the largest settlement in the Westfjords. Its airstrip, across the water from the town, could boast of a terminal building. In contrast, the airstrip at Suðureyri offered nothing more than windswept gravel on a hillside. "There is a tunnel here too," pointed Guðmundur proudly. "You see the entrance there? It goes right under the mountain, all the way to Súðavík in the next fjord. It is more than twenty kilometers long."

Historically, winter conditions isolated the settlements in the Westfjords for part of the year. Even in the best of weather they were only accessible by horse or foot, via narrow tracks over mountain passes. I could imagine how people might have felt

Suðureyri

trapped; closeted together in small communities with no way out for months on end. Isolation must have fostered petty tensions. By account, Ísafjörður suffered from such.

Ísafjörður, then known as Kirkjuból, is known for the most notorious witch trial in Icelandic history. It was in 1654. Trials and executions for witchcraft were at a peak in Scandinavia, despite a decline elsewhere in Europe. The pastor, Jón Magnússon, accused two singers in his church choir, a father and son, of making him unwell. He accused the son of also casting spells to give a local girl flatulence.

Both father and son, found guilty of sorcery, died, burned at the stake. All they owned went to the pastor, in compensation for the ill-health they had caused.

When burning the two men to death did not restore his well-being, the pastor accused the daughter of witchcraft too. Unlike the first, this case went to trial at Þingvellir, in the southwest of Iceland, where the court freed the woman. The court exonerated her and awarded her the pastor's belongings in compensation for wrongful persecution, scant remedy for the loss of her father and brother.

Witch trials continued, with many of the accused banished to Hornstrandir, the far northwestern part of Iceland, where we should head next if the weather would only ease.

After the plane landed, we waited beside the car until the dark-haired Parentis strode out from the airport terminal. With her, she had baby Hermann and her cousin Pieter, and numerous bags.

We spent the evening with Guðmundur, Parentis, Gunnur and Óskar. "There's a dance again tonight," they informed us. When Geoff perked up, they added, "There's nobody there yet. It's not worth going till late."

I am morbidly shy to dance, so I thought I would stay behind. Later, when everyone left to see how things were going, Geoff went so enthusiastically, I changed my mind and tagged along to watch.

At the austere venue, I kept my back to the wall and tried to go unnoticed, an impossibility as a newcomer to a small tightknit community. Geoff on the other hand, beaming with pleasure, danced away. He looked happy until a woman, appearing a little drunk, glued herself to him. Later, she looked as if she would devour him.

As the night developed, Geoff's expression drained from lighthearted dancing joy into concern, through discomfort into alarm. When he saw us leaving, he disentangled himself in a hurry and ran to catch up. His new friend followed us home, where Parentis took pity on Geoff and refused to let her in.

Suðureyri remained in sunny stillness, while beyond the fjord the gale continued to rage. Guðmundur tuned his radio to hear the updated forecast from the lighthouse, and to listen to the pilot of a Twin Otter plane that landed on the cliff runway. "It's not good," he summarized. "So, would you like to kayak across the fjord with me?" He smiled hopefully.

That seemed a fun idea. Norðureyri juts out as a low point of land, mirroring Suðureyri at the other end of the same terminal moraine. Crossing the gap of less than a half-mile, we landed to explore. What remained of an abandoned farm stood there. All the buildings were gone except for one, built with a solid concrete block in the shape of the bow of a ship facing up the hill.

"That's to cleave the snow," Guðmundur explained. "When we get winter blizzards, the north winds keep the mountains to the south clear of snow. We don't get much build-up above Suðureyri. It's different on this side. The wind eddies and drops all the snow along that edge up there." He pointed up the steep craggy slope to the ridge two thousand feet above. "It can get very deep, and sometimes it falls. Avalanches swept away all the other buildings from here.

Suðureyri

"In one avalanche, so much snow fell into the fjord, from this side, it caused a huge wave that destroyed the buildings all along the far shore in Suðureyri." I looked across at the town, realizing its vulnerability on that low spit.

We walked along the shore and cut uphill, soon finding it so steep we had to climb. The rock proved more unstable than it appeared, for a handhold came away. Above the others, holding the rock so it would not fall, I lost my balance. Bruised, and surprised at how friable the rock was, I guessed the reason for the ubiquitous scree slopes. I thought better of continuing to the top.

Waves refracted around the low exposed shore of Norðureyri, shaping up quite nicely, so I began to teach Guðmundur how to catch them to surf. Later, to satisfy his curiosity, I demonstrated a roll. One was enough for me. The water gushing inside my jacket felt so icy that, when he asked for another, I politely declined. Ready to stop, it did not take us long to cross the narrow strait back to Suðureyri, our empty kayaks light and lively, jittering on the water.

After dining with Guðmundur and Parentis, we walked to the dancehall to watch a film projected onto a roll-down screen propped at the front of the now movie-theater. The film was *Towering Inferno,* with Danish subtitles. A massive reel-to-reel projector balanced on a stand, amid the audience on folding chairs. It whirred loudly as the sprockets latched crisply into the spooling acetate. The cooling-fan blew air that smelled of burning dust, adding olfactory realism to the flames on screen. Faced with the drama, I could not help yearning for a little beach fire in the open air with a breeze from the sea. Towering Inferno was a world away from our experience of Iceland. We had to get going.

TWIN OTTER LANDS AT SUÐUREYRI.

GUÐMUNDUR TRIES THE VYNECK, SUÐUREYRI.

23
Rounding Horn

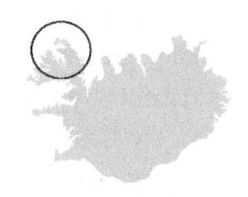

When the weather settled, it was time to move on. Parentis gifted us each a wonderful wool sweater. She had knitted both in an Icelandic circular yoke pattern using lightweight, loosely twisted, Icelandic wool. Wonderfully soft and cozy, the yarn seemed as if it should unravel and fall apart, but it did not. Like the balls of duck down we sometimes caught as they rolled along the beaches in the wind, such lightweight wool felt amazingly warm.

For the main body and arms of each sweater she had used a single color. The patterns in the yoke, and around the cuffs and bottom of the sweater, were in contrasting natural colors. I felt humbled. I loved meeting these people and would treasure the sweater as a reminder.

We lingered sadly over a last cup of coffee before paddling down the narrow fjord toward the open ocean. Someone waved from the lighthouse on the head and then we were on our own, creeping north against the cliffs, freed at last by the weather, silent in thought, lonely.

Five miles or less farther on, the view opened to the east, until we could see into the fjord, or *deep*; Ísafjarðardjúp. There, a fisherman in a small open boat stopped alongside to greet us.

Iceland by Kayak

Although he spoke no English, he tried to explain something by pointing, grinning, and repeating himself in Icelandic. We finally figured out what he meant. The open water distance to Ritur, north across the fjord entrance, was about the same as the distance to Neðri-Arnardalur, near the town Ísafjörður. We could see the latter about twelve miles away, while fog hid Ritur.

It seems the current here flows north during both flood and ebb, so we chose to cross to Ritur. Thanking the fisherman, we aimed on a compass bearing, presently reaching land amid clouds of kittiwakes, fulmars, razorbills, and guillemots. A fishy, tangy, smell came from the white-streaked water. Auks bobbed in groups, clucking, gargling, and ducking underwater. Lines of bubbles streamed behind them, like wispy contrails, as their sleek bodies rocketed, crisscrossing under our kayaks.

We cut across the next bay, about five miles to Straumnes, *stream nose or race point*, where despite the implication we saw no sign of fast current. Neither did the magnetic anomaly reported by the Admiralty Pilot, of ten degrees west, prove consequential. Somewhere in the cloud, atop the headland, stood the remains of the United States Air Force, general surveillance, radar station. Built between 1953 and 1956, it worked in conjunction with the south coast radar stations we passed at Reykjanes, and before that at Stokksnes. NATO run, it used to detect Russian planes, passing through the gap between Greenland and Iceland, until it ceased operations in 1961.

It was still foggy when we pulled into Rekavík, *drift bay*, to the east of the point, and made a landing onto an ankle-twisting stony beach. The chill damp clung in the air, on a breeze that shivered around carrying the cries of seabirds amid the shore-break clatter of tumbling stones.

The reason for the name, *drift bay*, appeared obvious. Driftwood buried much of the beach: huge logs, the majority felled, others uprooted. The trunks lay, stripped bare of bark, their ends blunt, and impact rounded. Smashed end-grain hung frayed and splayed. Most glistened, silky white, bone-like. Others in the damp glowed reddish. The sea had thrown them up along the top of the beach, jumbled one upon another, some buried, others half-hidden by stones and boulders. These logs did not come from around here. They had drifted to Rekavík.

As it happens, drift logs cannot float for long because they soak up water. Birch can float for a maximum of six months before becoming waterlogged and sinking. Pine and larch can float for up to ten months, and spruce for up to seventeen months before sinking. Most logs thrown ashore along Iceland's northern beaches are pine and larch, and that presents a puzzle. Ocean currents do not run fast enough to float them from any forested areas to Iceland.

Scientists learn the origin of logs by studying and comparing growth rings. In Iceland, even on the south coast, they have detected no logs from America. The Gulf Stream can and does carry American timber across the Atlantic, as far as the Azores. Somehow, the circulating current, slowing and dividing as it twists north and finally west, cannot carry logs far enough toward Iceland before they sink. Either that, or it whisks them off in a different direction.

Iceland's drift logs instead come from the Russian White Sea area, and from the Yenisey River basin in Siberia. They wash downstream from logging zones when the river ice breaks up. Floating free, they would sink before they completed the five-years-long, or so, journey on the ocean currents to Iceland. Instead, within months, they get frozen into the pack ice.

A part of the pack ice, carrying logs, joins the Transpolar Current to travel as far as the Fram Strait, between Svalbard and Greenland. From there the East Greenland Current takes over, flowing south. In cold years, pack ice carries these logs all the way to northern Iceland, accounting for the greater deposits in heavy ice years, such as in the late 1960s.

Mostly, the pack ice melts before it reaches Iceland, freeing the logs, a percentage of which float the last miles to wash ashore. Without the pack ice, Iceland would have no drift logs. I saw how most here lay jumbled well beyond the reach of the tide, thrown high by winter storms.

It was a wild spot. The beach formed a bar, a half-mile long and up to a hundred yards wide, across the mouth of a shallow fjord, trapping a lagoon that blurred into fog. In the other direction, the beach looked out over the Arctic Ocean. A low ceiling of cloud hemmed us in, dimming the day.

Geoff wasted no time before collecting sticks. He quickly set a roaring fire, prudently distant from the main piles of driftwood. Once we had warmed ourselves, we cleared an area of beach nearby, dragging away driftwood, and rolling aside the heaviest boulders, leveling a shelf wide enough for our tent.

As we worked, clouds of midges tormented us. Frantically swatting, it heartened me to realize that, unlike the midges I had suffered in Scotland, these did not bite. The tent set, we walked along the beach to find fresh water from a stream, hauling back a metal disk, the rusted-off end of a large oil drum, we spotted on the way. This steel plate, propped up over the edge of the fire, made a useful stovetop.

The fog gently lifted to reveal mountain ridges which, rising to fifteen hundred feet, flanked the bay and lagoon on both sides.

"What an incredible change! I've so looked forward to getting up into this area," I confessed. "Apart from the south coast sands, this must be the most remote part of Iceland; no connecting roads, close to the Arctic Circle." Looking down at the sticks, twigs, and beams that lay scattered at my feet, I added, "I'm glad they delivered the wood for us."

Rounding Horn

"Yeah! It should be enough to last us a day or two at a pinch, don't you think?" joked Geoff, beaming at the abundance. He seemed as proud as a schoolkid. Beside him, his fire blazed like a beacon.

Usually, I prefer little fires, small enough to cook on without getting scorched. Here, Geoff's bonfire felt wonderfully powerful. The skin on my face tightened in the radiant heat as I circled, warily avoiding the sparks that spit and spiraled. Gas hissed loudly from a hole, flaring like an acetylene jet. The smell of driftwood smoke, dry in my sinuses, differed somehow from woodland bonfire smoke. Was the wood drier, or did sea salt make it smell different? Occasionally the rifle-crack of a stone heated too fast, or the soft settling of logs scrunching down into embers, startled me.

Turning to cool my face and warm my back, I noticed what looked like a small grey naval vessel passing stealthily offshore. Icelandic, I figured, from its proximity to land, and if so, a coastguard vessel.

Iceland, with no navy, supports a handful of defensive coastguard patrol boats like this. They figured prominently in the cod wars. As this one headed west, I watched its progress until it slipped from sight behind the cliff of Straumnes.

I opened my map. I might describe Iceland's northwest peninsula as an equilateral triangle pointing north, its southern base along Breiðafjörður. It joins the main body of Iceland at a narrow wrist of land at the southeast corner of the triangle.

A series of gashes, the fjords we passed, cut southeast from the northwest side of the triangle. The longest of these, Ísafjarðardjúp, which we crossed today, isolates the long and mostly uninhabited peninsula where we stood. This peninsula forms the blunt northern point and turns into the north-east facing side of the triangle.

Iceland by Kayak

On the map, I thought this north coast part looked like a multi-legged creature with a narrow body. There were steep ridges rising to more than two thousand feet from sea level, dropping each side as steep mountainsides or precipitous cliffs into valleys or ocean.

The nearest connected road to here was thirty miles away down the peninsula, beyond the more than three-thousand-feet-high Drangajökull ice cap. From where I stood, looking straight out from the beach over the dark northern sea, the horizon hid only the polar ice cap. To my left, 180 miles beyond the thousand-feet-high ridge, in the direction the patrol boat passed, lay Greenland.

At this northwest tip of Iceland, we stood at a corner. From here, we would turn toward the rising sun. Tomorrow, if all went well, we planned to reach Hornbjarg, one of the most remote inhabited spots in Iceland. At Hornbjarg stood a lighthouse, its keeper, Johann, a well-known figure. I learned that he owned the largest private collection of books in the country, quite an achievement in a land of prolific readers and book owners. So many people we met knew about Johann and urged us to visit him, I looked forward to the opportunity.

Back on the water next morning, my mood was unaccountably happy. This place seemed different. It was wild and empty, but it was more than just that. Iceland had shown us myriad places with mountains, and cliffs, and steep-sided valleys, so why did I find it especially beautiful and dramatic here? What was extra? Was it the different light, on the north side of the cliffs?

Balancing my elation, Geoff seemed somber, and I wondered why. "The water feels really cold," he explained at last.

"Yes, we must have picked up the East Greenland current. It is colder."

In truth, the whitewater rivers in Wales, where I paddled in winter, were frequently colder than this. Sometimes ice encased the rocks and riverbank foliage. When I stopped below rapids my beard froze to my jacket. Those short winter descents were

Rounding Horn

bitingly cold. What we experienced here, wet hands for hours on end in a chilly wind, proved differently harsh. A small drop in water temperature was noticeable.

I ignored the cold. The coast looked spectacular. The cliffs, which often soared dizzyingly until low clouds obscured the tops, were steep and horizontally layered. Sometimes pinnacles and rock towers adorned the edges, snagging the clouds, threaded between by whisps of haze. Around the cliffs, the air seethed with seabirds, flying over us at every level. The highest, a thousand or more feet up, looked tiny, like swarming midges on a summer evening.

Massive sea stacks, like giant gravestones, stood dwarfed by the dark cliffs. Contrasting the shadow, white horizontal bands marked where birds nested in their millions. The soaring cliffs stood bold in grim majesty against the Arctic Ocean that beat at their feet.

Between headlands, broad green valleys swept down toward the shore, wave-trimmed, or crumbled into low cliffs above a narrow boulder foreshore. Sometimes I saw the gleam of light on sandy flats.

On the water beneath the cliffs, rafts of guillemots, razorbills, and puffins, undulated with the swell, dipping under breaking crests, skittering away, or ducking when we passed close. Their growling and clucking added a layer of cacophony to the background rumble and report of swells hitting the cliffs. Plumes of spray blossomed up before white frothy trails clawed back down the rock. As the swells rebounded, the water sucking down to expose ever more cliff, the reflected wave hit the next oncoming swell and erupted into chaos, hiding the gaping trough behind.

Suddenly, I spotted a single tiny bird, which looked like a shore bird, spinning in tight circles amid the chaotic water. It looked too small and trim to be a seabird. While the guillemots looked like sturdy torpedoes, this bird sat daintily upright and gyrated as if showing off. We had seen others like this before, usually alone, close to shore by the rocks, and wondered what they were. This one looked as incongruous as a puffin in a city. Undisturbed by my approach to within feet, it continued to spin while I sat and watched. What was it?

"I've no idea," admitted Geoff, before we left it to its sorcery. "You have to admit, it is cute."

It would be another twenty miles to Horn, where the sea looked wild even from afar. Hoping to avoid the tide race which forms there, we approached during the ebb, expecting the ocean current to balance out the opposing west-going tide. So, the turbulent sea ahead puzzled us, until we reached the point and hit the wind funneling around the huge cliffs. It was time to pit our strength against the buffeting gale, a slow grueling grind, pitching into chaotic rebounding waves that leaped and burst into billowing haystacks of spray.

It grew ever darker and oppressive, close beneath the cliff, where we pushed on against an ever-twisting wind. The sullen wall above us, already towering a thousand feet above our heads, soared nearby to more than fifteen hundred feet. Fulmars spun past, standing on a wingtip. Changing direction as quick as a flick, a shift from one wingtip to the other and the bird was gone, snap, blown away. Puffins and guillemots sped like bullets.

Paddle stroke after paddle stroke I pushed on, one blade grasping the water while the other quivered in the gusts. As the waves juddered from rebounds, my blade, one moment anchored deep in water, next clutched only air. I was on a knife edge of balance, constantly alarmed and unsteady. Thud. Boom. Crash. The percussion of breakers, thundering into foam against the rock, seemed to vibrate in my bones.

All around, the water hissed and flew like rain. Breaking crests hurled waves of spray, blinding us. White clods of spume hurtled by like ghost birds or tiny clouds. I clenched my paddle

Rounding Horn

tight, glanced with salt-stung eyes to check on Geoff, and forged on.

The wind did not relent until we reached and rounded the point. Gradually, it eased, the funneling less intense, the gusts sporadic. I began to relax. "Sheesh!" I gasped, exhausted and relieved. "That was incredible! But look, there's the lighthouse." The lighthouse and other buildings of Hornbjarg stood up on the vast clifftop ahead. Miniscule in contrast to the landscape, it was the color and the regular shapes that had caught my eye.

We passed a bowl-shaped bay, and three ominous tombstone-shaped giant slabs of rock facing out to sea, to pause offshore the lighthouse on the cliff. The sea was wild and chaotic between us and the shore, which presented no promising place to land. The swell looked too daunting. Huge waves dumped across rocky shoals and pounded over hidden ledges.

"Look! There are steps down the cliff, so this must be the right place to land." Yet I could not see a safe way to approach. Everywhere, breakers pounded over reefs that lay hidden from our view.

"You know, I don't like the look of this," I confessed. "We'll never get in there. The only thing is, the next stretch of coast all faces the same direction, so we really ought to try."

I was torn. My sights were set on Hornbjarg, for we would surely not have another chance to meet Johann. This remote point was accessible only by boat, or by walking for days along the peninsula. In the end, I concluded, we must play safe and paddle past, however unappealing that decision. Landing looked too hazardous.

We hung back and watched. The bay was a mass of creaming breakers. For a moment I thought I saw a route. In the next, a swell reared up. Pitching out toward the beach, it tubed, collapsing with a deep *whomp*. There must be a hidden ledge there. My adrenaline raced.

Iceland by Kayak

"That looks the easiest option over there," suggested Geoff. Moments later, a huge swell reared up and curled right across where he pointed. "Oops," he said.

"Look, there's someone up there." He was looking to the right, up the cliff, and I followed his gaze. Figures waved from the cliff top. When we waved back, they pointed, signaling us south. "They want us to go that way." We complied until they waved us to stop, beckoning us in. "Really? Do you think they know what they're doing?"

"I'm sure they can see better from up there. We should trust them. Shall we try it?"

I was scared. To either side of us, the swells gathered themselves up and thundered across the rocks. The air was full of misty spray. I could feel the percussive noise of each breaker through the seat of my kayak, as much as I could hear the roar. We were in a narrow gap of deeper water, a passage. Following the directions from the cliff, we turned. Keeping a low cliff on one side, surf breaking over rocks to the other, we hurried toward a steep beach, pausing for a moment to time our landings, on the backs of waves, onto the boulders.

"Phew! That was too exciting!" I acknowledged, hauling my kayak clear, relieved to be ashore. Tensions released, I felt utterly drained.

Two men, descending to meet us, helped us carry the kayaks high up the beach. They said they were here to renovate the lighthouse buildings, adding, "They date from 1930."

We followed them up the steep steps to the clifftop where we were surprised to find a young and pretty woman waiting for us. She introduced herself as Olga, a relative of Johann the lighthouse keeper, although I did not grasp clearly whether she was his niece, or his cousin. It was Olga who had spotted us approaching and had the good sense to guide us in.

Rounding Horn

"Look, you can see the way quite well from up here," she pointed out. "You would not see it from out there. I saw you stop and realized you might appreciate help." She was right.

"We're so glad you did. I don't think we would have landed otherwise. We couldn't see a safe way in."

"I didn't expect to see you coming along today," she said. "Now, come and have a shower and I'll make coffee."

As we walked, she chatted, "Johann will be sorry to have missed you. The lighthouse was due for repair and repainting, so yesterday the authorities lifted him by helicopter to a coastguard boat. He'll catch a plane from Ísafjörður to Reykjavík for a holiday while all this is going on." I was disappointed. I had really looked forward to meeting him, and we missed him by just one day. Was he on the ship I saw yesterday? Probably.

"Anyway," Olga continued, "I'm looking after the place for him while he's gone," she turned with a wide smile. "You'll have to make do with me. Is that okay?"

Showered clean, seated in front of a tall flask of coffee and a generous spread of food, encouraged to "eat as much as you can," we were in heaven. Afterwards, I offered to wash the dishes. Since the renovation team was currently refitting the kitchen, I took the dishes to the bath.

Now, Olga gave us a tour. "This is Johann's library. He has one of the largest private collections in Iceland." The books were beautifully bound, neatly shelved all around. "He has another roomful downstairs." Had he read them all, I wondered? Was his collecting an obsession? Did he have specific interests that themed what he collected? With so many questions I would have asked him, I suddenly had a thought. "I wonder, does he have any bird books?"

"Yes, of course! He has all kinds of books. Why?"

Iceland by Kayak

"I saw a little bird today, on the water, spinning around on the surface as if it had lost something. It was a tiny dainty bird, not at all afraid. I have no idea what it was. It was new to me."

"Let's see," she said and soon had a large, illustrated book open at the desk. "It's in Icelandic, of course, but the Latin names will be the same."

Never skilled at Latin, I remember only a handful of birds by their Latin names. When I saw a depiction of the bird I had seen, I jotted the name in my notebook: *Phalaropus lobatus*.

We sat talking into the night until Olga stood up, pointing out that she would need to be up early to feed the workers. "You may sleep up here," she invited. We followed her upstairs. "I'm afraid we don't have a spare bed; can you sleep on this carpet? You're probably used to sleeping on the ground?"

I looked at the thick rug. "This will be luxury, thank you! Only, it's so hot up here. I wonder, would you mind turning the heating down a little, please?"

Next morning, the sound of hammering downstairs woke me. Work had already started. The comfort of the cushioned floor made me reluctant to get up, but I joined Geoff in the kitchen where Olga prepared us a hearty breakfast. When she invited us to write a short account in the lighthouse *Gestabók,* reading back through earlier entries I recognized a name.

A group from Castlebrae High School in Edinburgh under Chris Sugden, someone I had met before, had left an account of an expedition they had made here overland. From his account I learned that the common name for the tiny bird I had identified as *Phalaropus lobatus,* was the red necked phalarope.

Seeing my interest in birds, Olga explained, "Nowadays hikers often come here with binoculars, just to see the birds. We get a lot of birdwatchers. In the past it was different. People came to collect eggs, Svartfugl eggs. Is that guillemot? The birds with the black backs. Anyway, there is more than one kind. They went

Rounding Horn

over the cliffs on the end of a rope. When they were ready, the others would haul them back up again. They would take two or three hundred eggs each, to sell in Ísafjörður or Reykjavík.

"One time when a youth went over the cliff to collect eggs, they hauled him back up to find the whole of his lower jaw missing. He was dead of course. Falling rocks often killed egg-collectors. There are so many birds, and the cliffs are frost shattered. It was very hazardous."

On a lighter note, she continued, "Yesterday, forty people visited the lighthouse by foot. Forty, in just one day, and the nearest road is five days walk away. That was busy."

"Where did they all come from?" I asked.

"They were mostly Icelanders, on holiday."

Not a bad place for a holiday, I thought, reluctant to leave here, or her. I would have liked to have stayed longer, even until Johann returned, to meet him too, but we had to move on.

The kayaks were at the water's edge and loaded ready to launch when a group of hikers came down to the beach to watch. Icelandic, they had walked about fifteen miles per day to get here. For all its isolation, or because of it, this lighthouse is a destination.

It was easier to launch than it had been to land. Facing the waves rather than following them, the wind having eased, and the sea settled, we could easily see the safe route from the beach. Heading south, we watched the sky, overcast with enough low clouds to smother the view. Occasionally the clouds broke open in places, offering a teasing glimpse of mountains. The gullies and corries, on the mountainsides where the sun had not reached, were still full of snow. Suddenly, we spotted the top of Drangajökull ice cap, shining up high.

My wrist ached from yesterday. In the conditions by the cliff, trying to keep balance, I had undoubtably gripped too hard with cold hands. Today, wearing a neoprene band to keep my wrist warmer, I held my paddle as lightly as possible and tried to avoid bending my wrist.

It irked me, reminding me of a late season solo journey I had made two years back, paddling by day and sleeping out on the beaches. Exploring the Isle of Wight and part of the south coast of England I had grown impatient, frustrated by the shortness of the winter days. Defiantly, I had paddled long into the night, aiming along the narrow moonwake. In retrospect, I understood, balancing on the lurching waves in the dark, I had gripped my paddle too tightly with cold hands for too long. Waking next morning, where I had slept on the beach beside a fishing boat at Worthing, my wrist was swollen and very painful.

Days later, I sought help. "Tino-synovitis. Inflammation of the tendons," my doctor diagnosed brusquely, adding that he could do nothing about it until the swelling subsided. "Don't use it. Come back in a few days, when the swelling's gone down, and I'll put it in a plaster cast for you."

I was crestfallen. A cast would prevent me from paddling, so I never went back to see him. But soldiering on as normal without a cast, my wrist never seemed to heal. Now, that same aching pain had returned, and I worried it might prevent me completing the trip.

We carried a packed lunch. Geoff had helped Olga to prepare it in the morning while I washed dishes. It had looked so appetizing, carrying it was a torment. The temptation to stop right away to eat was almost unbearable. We had to either agree to paddle a certain distance before stopping, or to decide on the earliest time we would allow ourselves to eat.

At last, we chose a landing beneath a dark basalt wall, one of a group of sea stacks standing like books on end, each bearing a crazy honeycomb pattern on the cover. Every irregular hexagon, the end of a large block of rock, fit perfectly against the next, together forming these stacks: tall, broad, and thin. There we sat

Rounding Horn

to unwrap our giant sandwiches, feeling so grateful to Olga, and wishing we could eat like that every day.

The rock formations here, at Drangatangi, *stacks point*, made me think of mud. Mud, when it dries in a puddle, shrinks, and tends to crack into polygonal plates. It dries from the exposed surface, cracking downward. The horizontal surface plates, with cracks between them, gradually form into vertical columns. More accurately, not completely vertical, for as a puddle dries, it first shrinks down into a bowl shape. The blocks, or columns of mud between the cracks, converge toward that dished surface.

Basalt, unlike mud, begins as molten lava, which shrinks and cracks as it cools. I assume the columns usually point up in the direction of cooling, toward the earth's surface, in vertical columns away from the heat of the earth's mantle, just as the columns in mud point up to the drying surface. If that is so, the cooling surface here must have been to one side of the lava, not toward the earth's surface. Either that, or the land here has toppled ninety degrees. The garlic-press type of formation we frequently saw, where columns appear to spray out in all directions, forms when water affects the cooling.

Here at Drangatangi, the horizontal columns looked like the workmanship of Inca stonemasons, each block fitting perfectly. In places, a block had fallen out to leave a hole. Elsewhere, more pieces had fallen, creating flower-shaped recesses, the perfectly nested ends of the shortened columns tucked inside.

The temperature began to fall, and the sky darkened. It was time to move on, and warm up again. Olga had told us yesterday's temperature reached seven degrees Centigrade, dropping to five (41°F) overnight. Somehow, it seemed colder than that here, probably because we had stopped moving and were wet.

Day still blurred into night without getting dark, so it surprised me later when I saw the almost full moon, bright in the sky. It was the first time I had been aware of it in a long time. The nights had grown darker so imperceptibly, week by week, I had not noticed any change. Yet tonight the moon shone clearly. The sun set around ten-thirty and would rise at four in the morning, with this twilight between. Earlier in the trip, the sun stayed up until as late as midnight, returning by three in the morning. It had always been light.

We crossed narrow fjords between razorback ridges, and passed one or two small farms, to stop near Gjögur on Reykjarnes. The shore looking soft, I sped toward it confidently, grimacing at the shriek, like fingernails on a blackboard, when the fiberglass hull ran across something sharp in the mud.

With more than forty miles behind us, in weary satisfaction we lifted the kayaks onto the grass and set up our tent toward the view. Angular peaks to the south stood a couple of miles away across Reykjarfjörður. A trickling stream supplied us with clear, sweet water. Soon, replete with hot food, it was time to sleep. Tomorrow, or so we planned, we would set off east across the open water of Húnaflói. We had not accounted for the weather.

Rounding Horn

GJÖGUR.

Iceland by Kayak

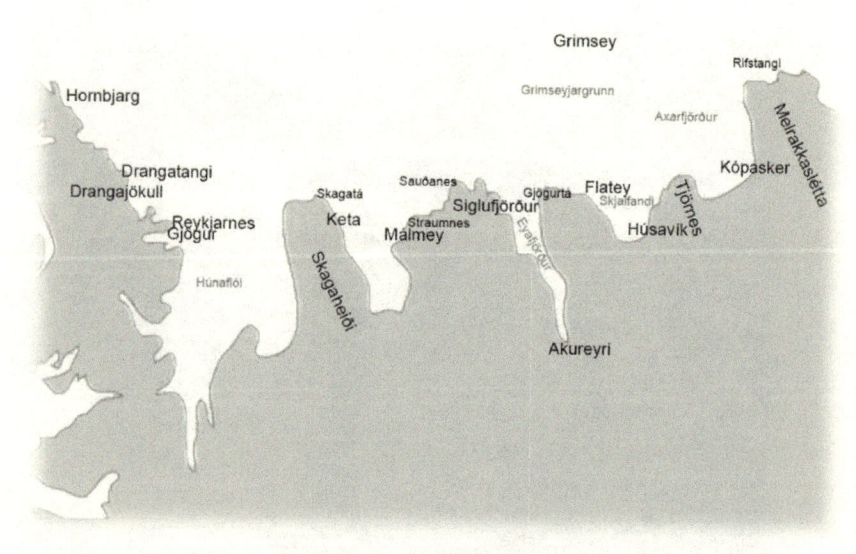

MAP 8. NORTH ICELAND.

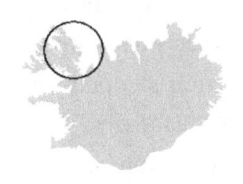

24
Gjögur

The flapping tent woke us. Whitecaps embroidered the emerald bay, into which flocks of screaming terns took turns diving for slender fish which glistened silver in their blood-red beaks. Waves tumbled onto the rocks. "Not the best conditions for crossing Húnaflói really," Geoff considered, his resignation voicing what I had already concluded. To cross Húnaflói would entail twenty-five miles of open water against this easterly wind. Paddling around the bay would be at least three times that distance. Or we could wait.

"No, not really practical," I agreed. With that decision made, we zipped up the tent and went back to sleep.

The wind increased throughout the day, bringing intense squalls of rain under menacing, monster dark clouds. I piled more and bigger stones around the tent, and onto the clothes we had pinned to the ground to dry, although it was debatable whether anything would dry between the showers.

As I cooked, stirring canned sardines into rice, I gazed across at the mountains, which climbed steeply in layers from the water across the fjord. Like so many mountains here, the base from the water was a green slope, steepening as it rose to exposed crags, which stepped up, alternating between rocky ramps, and near-vertical faces. They were reminiscent of the crumbling edges on a multilayer cake.

The tops thrust up, spiky, with sharp-edged ridges. This fjord, unlike Húnaflói, was only about two miles across, so the mountains facing us stood intimately clear. Each steely rain shower gave way to patches of sunshine that spot lit one peak at a time, leaving others in shady silhouette. The rain often shadowed into view as dark, gauzy veils creeping across the mountains, before each squall drove an escort of spray across the fjord.

I sat to watch the terns' wheel and hover in the wind. They shrieked and bickered together in a cloud, chiding and jostling. Their sharp shapes darted down to snatch morsels from the tumbling crests, sculling forcefully back up. Frosty plumage, brilliant in the sun, flashed against the backdrop of dark mountains and water. It was invigorating to watch and listen, but also relaxing. I melded into the scene, reassured that there was no hurry to leave.

Exploring from camp we came across an airstrip, a short distance inland, just a leveled unsurfaced runway. We also found neatly stacked sawn logs; the broad trunks cut into about six-foot lengths. But we hastened back at the first spits of rain before the next deluge.

In the evening, Geoff walked off toward the hamlet, Gjögur. After a while I followed in the same general direction, meandering first along the beach, then inland. I passed small hayfields, each enclosed by wire netting hung between driftwood fenceposts.

Open boats lay on the grass above a gravelly beach. Nearby, fish hung drying on a rack under a pitched roof: a shed without walls. I stopped to look more closely. I could only guess at what

Gjögur

else swung in the wind beside the fish. Gruesome, it looked like chunks of flesh.

Ahead was a small farm. Intending to ask for a weather forecast, I approached, walking between tiny driftwood-and-chicken-wire enclosures. The place was unruly, and I grew ill at ease. Piles of sodden wool fleeces lay out on the ground in the open. Heading for the house, I passed a small barn with corrugated-sheet siding.

Close by, two young women raked cut grass, piling it into wheelbarrows. They laughed freely at my antics when I tried to convey that I wanted a weather forecast. They beckoned, and I followed them to a workshop, opposite the house, where they stepped aside to let me pass into semidarkness. Feeling unaccountably apprehensive, I hesitated before entering.

I jumped, alarmed, scared by a movement in the shadows. There was a figure, almost hidden in the gloom. The shape moved again, and as my eyes strained in the dim light, I saw it was a broad elderly man, in dirty brown clothing and flat cloth cap, bent over almost double. He looked straight at me with intelligent, seeking eyes. From his nostrils ran a thick ooze of blood, or something similar, which he made no attempt to wipe away. I wondered whether this was from a recent accident or was a permanent feature of his face.

When I asked if he knew the weather forecast, the old man's eyes so fixed me, I did not notice there were other men there too. Startling me, one questioned rapidly from the shadows in Icelandic. Pausing for a moment for a response, and getting none, he crept closer as he continued to speak.

My eyes adjusted; I could see gaps where the old man was missing teeth, his remaining teeth dark, his chin thick with beard stubble. Commanding, bright, and relaxed, he seemed to be the one in charge.

Iceland by Kayak

The others kept quiet. One with a bushy beard worked under the hood of a Land Rover, looking out from time to time with an amused expression. Another, with wild eyes and a bristly chin, stood half-bent over, echoing the stance of the older man.

A taller young man, lacking the stiffness of movement of the others, leaned against the wall, and watched me. He casually got out a pipe, filled it, and lit it. Holding it with its long stem in the air, he smiled at me, amused at my reaction to the barrage of questions from the old man, or at the old man's questions. I felt intimidated.

As I tried to explain what I wanted, the men gathering around me, one of the women reappeared carrying a young child. The other, near my own age, followed. They were both smiling. I had a way out.

Doubting anyone understood a word I said, I thanked them all and turned to leave.

"Wait!" It was the tall man. He explained in broken English that someone was already on his way here, someone who could speak better English.

As I waited, the old man pulled from his pocket what looked like a small oil can, holding it out, offering it for me to take. Everyone watched attentively. I gestured, "No thanks," and watched as he turned and carefully blew his nose on a dark cloth. He poured a mound of powder onto his wrist from the can and sniffed it up into one of his huge nostrils. He followed with the other nostril. I could scarcely believe what I saw. I realized at once that what I had mistaken for blood, between his nose and lip, was the stain of snuff.

After a pause, he continued talking at me until a thin man, with glasses and a bold tartan *tam-o-shanter,* came striding in, followed at his heels by a petite Chinese woman.

"Hello," he greeted me, "I can speak English if you would like me to translate."

Gjögur

I explained how I was only looking for a weather forecast. "There are two of us, with kayaks, paddling around Iceland. We're camping just along the shore, over that way," I pointed. "We need calmer weather to paddle."

"The weather should stay like this for the next twenty-four hours," he told me. "It will come from the southeast and might increase a little." It was already blowing about force six, so that did not look promising for a departure in the morning. "Come back here to the farm tomorrow, at noon, to phone me. I'll tell you what the new forecast is."

Thanking everyone again, I left self-consciously, retracing my path between the fences. A shout called me back, a man calling. He gestured, offering me a ride in his Volkswagen Beetle. The old man, and a young boy, climbed aboard with us, and we bounced off toward the lighthouse and the airstrip. From there, I was able to point out our little orange tent across the fields. With no easy closer approach, they dropped me off to walk the last stretch.

Geoff was already at the tent. "I went to the farm for a weather forecast," I explained, when he asked. "It was challenging. Nobody spoke English at first, but I learned that the wind should continue to blow like this, and stronger. They said to come back tomorrow for the midday forecast. How about you? Where did you go?"

"I went over to the airstrip and helped a man paint a hut. I didn't have much luck talking either."

Next day, Friday, we walked to the farm together. The fleeces: white, black, grey, and ginger, yesterday on the ground, now hung over the fences to dry. I am familiar with the sheep raised in the south of England. There, the Southdown sheep from Sussex, Britain's oldest breed, and the Sussex Romney, are both burly, with coats of monotone creamy white. Icelandic sheep seem petite by comparison and come in a range of colors. Despite their dainty appearance, they must be hardy.

Iceland by Kayak

Originally, the early settlers brought sheep from Norway, of the same domestic stock as Norwegian Spælsau: northern, short-tailed sheep. Subsequently, these sheep have seen more than a thousand years of breeding and adaptation in isolation. The only breed of sheep isolated for longer is the Soay, on the remote Scottish islands of St Kilda. In the early days of Iceland's settlement, people kept sheep primarily for meat, also for wool and milk. Since cows were expensive to keep, until the 1940s people more commonly used sheep milk.

Before there were sheep here, trees covered forty percent or so of the land in Iceland. These downy birches, rowans, willows, and aspens had value as building timber and firewood. Few trees remained after the first three hundred years of occupation once settlers had finished clearing them away for farmland. Sheep have since kept the vegetation so well-trimmed that, by the 1950s, only one percent of the land had trees.

We stopped to look at the wooden boats: two open boats, and a third with a small cuddy. One open boat, twenty-five feet long, lay on the grass for repair or modification. New pieces of planking were already in place where the red and green paint ended on one side. Those planks still needed trimming, and there were other parts that also needed maintenance. The other boats jostled on the water, moored between ledges of rock. Three capstans nestled in the grass to haul them all ashore.

"What a nice little setup," approved Geoff. "A small farm, with fishing boats."

When we reached the farmhouse, the old man welcomed us into his kitchen, where everyone else had already crowded in. I introduced Geoff, and we learned the man's name, Axel. While the family vacated two stools for us to sit, Axel's wife cleared the table and set out clean mugs. Next, she brought a thermos flask of coffee, and cakes on a plate.

Gjögur

We sat at a table by the window. Across the room stood another table. A kitchen range and counter lined the third wall, while an enclosed ladder or steep staircase, opposite, led up to either the second floor or a loft. Aside from the odor of wool and leather, the room smelled a little unpleasantly of something I could not identify.

Axel was very cheerful and animated, so Geoff tried to communicate, yet despite his creativity he could not better his simple gestures, such as pointing at coffee, sticking his thumbs in the air, and smiling. Everything was awkward until the English-speaker arrived with news that the weather would be bad. It would stay that way for at least the next twenty-four hours.

When he left soon after, we finished our coffee and stood to follow. Axel stopped us. We must eat, he invited, so we joined the family for cod, and small, sweet-tasting, new potatoes boiled in their thin skins. To follow came a semolina pudding, with raisins and cinnamon.

In the middle of the table, throughout the meal, sat a lump of fatty looking jelly on a plate. Geoff had been eyeing it with curiosity and drew my attention to it with a quizzical look. I had no idea what it was. In the end he asked, "Is it tongue?" He pointed at the lump on the plate and opened his mouth to point at his tongue.

"Nay." *No.* There was a chuckle.

Geoff mimed drawing on the table, and someone hurried to fetch paper and a pencil. We took turns drawing a sheep, a cow, and a pig. Finally, one of the men brought a photo of himself with a seal and what might have been a dolphin he had caught. The photograph was in black and white, and was a little blurred, so it was difficult to see exactly what it showed. "Maybe it's seal blubber?" Geoff surmised, conferring with me. "Fishermen often shoot seals when they come near the boats, don't they?"

Eventually, Axel slid the dish across the table to Geoff to sample. Geoff took a small piece and popped it into his mouth. His face writhed, contorted in disgust, prompting hearty laughter from all the others. "Revolting! It's absolutely…" His face screwed up, speechless.

After seeing his reaction, no amount of his urging could persuade me to taste it. I did smell it, tentatively, recognizing it as the source of the unfamiliar stench I detected when we first stepped in here.[3]

After the meal ended with more coffee, Axel invited us to come back again at seven-thirty. We accepted, asking a favor in return. "Can we help you this afternoon?"

Stooks of hay stood in the nearby enclosures. Our task was to pull them apart to spread on the ground to dry. The whole family joined in this activity, so we learned by copying. Next, with the rakes held at an angle, we walked back and forth along the rectangles of hay, drawing the hay into long narrow lines. As we raked, we began to learn about family farm life, a little bit from each person in turn, as they became more confident speaking the English they knew.

The farm had two Lister engines, I presumed to generate electricity. It was the men who went fishing in the boats we had seen. The family either ate the catch fresh or hung the fish by their tails in the drying rack, the frame with the corrugated metal roof we saw near the farm. They occasionally hunted seals for meat.

They kept sheep, they said, mostly for wool, also for meat. They rounded up the sheep into the farm each winter to shelter inside, although in mild weather they let them out to graze close by and to forage along the shore. The family washed and dried all their raw wool here.

One of the women explained how Icelandic sheep had two types of wool: a water-resistant outer layer of thicker, coarse, long hair, *tog*, and closer to the body, a coat of thinner, softer, short

Gjögur

hair, *þel*. In rain, water sheds from the tog, keeping the insulating þel dry. But even when the whole coat is soaking wet, the wool insulates well.

The family washed the raw wool and separated it into the two wool types. Then, they carded it, and twisted it loosely into yarn for sweaters, hats, and gloves. The sweaters Geoff and I wore were of the soft *þel*. To knit such traditional circular yoke patterns required more than two knitting needles. Here, they never dyed the wool, instead using the colors from different sheep: black, white, ginger, and grey, for the decorative patterns.

Apart from sheep, the family kept nine chickens for eggs, and grew radishes, rhubarb, and potatoes, in the small enclosures.

Axel proudly showed us the silage he was making. The strong sweet smell was overpoweringly pungent. Would it smell less once fermentation was complete? But then again, if sheep forage for the pungent seaweed thrown ashore by storms, the strong smell was unlikely to be a deterrent.

Axel also explained how all the driftwood we see, piled high on the beaches, floated to here from Siberia. The logs, from the salt in the water, became almost impervious to rot, ideal for fenceposts. The men cut the logs into shorter lengths before splitting each piece, into narrower posts, by driving in wedges. Each cut section of log yields at least a half-dozen posts.

A single fencepost sells for three hundred Króna, so each post-length of log could make up to 1,800 Króna. A whole log could be worth four or five times more. In context, each postage stamp, to mail a postcard home to England, costs me forty Króna. I could not bring myself to confess our extravagant bonfires.

Axel led us to the beach where he pointed across the fjord to each mountain in turn, naming it and asking us to repeat the name. They were his friends, and he looked happy introducing us to them.

Later we raked the grass into mounds to load a small trailer behind a Jeep. Someone had secured the rusted hood with two lengths of string and replaced oil and gasoline filler caps with wooden stoppers. Despite its appearance, the Jeep ran well. However, my head banged against the cab roof as we lurched along the rutted track to the barn. We left everything there and took a break. Fortified with sausages and potatoes, we piled all the hay into the little barn before crowding into the kitchen again for coffee.

Axel, who considered our socks inadequate, gave us handknitted wool socks. I was very content. I could have happily stayed and helped him in return for food, for the rest of the summer. The pace of life, and the family atmosphere, seemed wonderfully relaxed.

Curious about the relationships between family members and friends, I learned that Axel and his wife had six children. Five lived here: Ölver, Ólafer, Jacob, Carmilla and Elfa. Johanna was in Reykjavík with her husband, while their son Ivar stayed here for the summer with his grandparents. They said it was common for children from Reykjavík to spend time in the country, with relatives, during the summer school break.

In turn, they asked me if I had siblings, was I married, and when was my birthday?

"Three siblings: that's two brothers and a sister, and no, I'm not married," I explained. Axel seemed pleased to hear that my birthday fell on the same day as Elfa's. At twenty-one years old, she was exactly three years younger than me.

The other men who came and left last night, and today, were friends and neighbors, not family members.

It was tiring trying to follow the conversation, in a crowd speaking Icelandic, even with bits of translation. After arranging to return the next day for a weather forecast, we made our way

Gjögur

back to the tent, my head awash with partially understood conversations, new experiences, and unorganized ideas.

Geoff woke me up at about seven o'clock the next morning. The sun shone, and the sea mirrored a cloudless sky. When Geoff returned to his sleeping bag, I cooked breakfast, wrote in my journal, and began to pack my kayak.

Today, was July 30th. We had one month left before I was due to resume work at Burwash. We ought to catch the ferry from Seyðisfjörður on 20th August, the sailing before this year's last. My quick calculation suggested we had 250 to 450 miles still to go, to reach Seyðisfjörður. The exact distance depended on how closely we followed the coast. Roughly measured, we had so far traveled 750 miles in forty-eight days, of which we had paddled on twenty-seven. Our average progress had been fifteen-and-a-half miles per day overall, or twenty-eight miles or so each day paddled.

Taking those averages, I tried to estimate how long it would take us to get to Seyðisfjörður, but my results ranged from nine to thirty days, depending on our exact route, and allowing for the weather. It was futile to second-guess our timing, our route either crossing bays or following the shore, or the weather, but with twenty days left to make the ferry, there was no reason not to be optimistic.

Geoff refreshed, we walked to the farm to get a weather forecast. Over a meal, we learned that the weather update was favorable. Reassured, we packed and launched, planning to make a short detour to show the family our kayaks, before turning east, across Húnaflói.

FLEECES DRYING, GJÖGUR.

AXEL AND HIS SON AT GJÖGUR.

25
Siglufjörður

Geoff seemed irritated. Was he frustrated, having waited for the favorable midday weather forecast at the farm, instead of leaving earlier, when we could have judged the conditions for ourselves? The forecast was clearly inaccurate. When we first aimed straight across the bay, Húnaflói, toward Skagi, *peninsula*, which lay hazy in the distance, conditions were ideal. Now, the wind made our crossing uncomfortable, whipping up waves enough to keep us soaked and chilled. The sea grew too awkward for us to paddle close together, making conversation impossible. We each suffered in our own little world.

Iceland's historic sea battle took place here, in Húnaflói, in the mid-thirteenth century. That was during a period of civil war which began in 1220 and continued for more than forty years. Local chieftains gathered their followers into armies, fighting each other for power, and for revenge. The Sturlunga family clan, which controlled western Iceland, the Westfjords in the northwest, and northeastern Iceland, was one of the most powerful clans at the time, so the period of wars is known as the *Age of Sturlungs*.

King Haaken IV of Norway wanted to unite Iceland under his own sovereignty, so he courted Iceland's most influential figures.

Iceland by Kayak

He invited the chieftain of the Sturlung clan, Snorri Sturluson, into his court. Snorri was a famous poet, historian, and politician, from Westfjords. But over a period of years, in which inter-clan battles ended the lives of multiple clan leaders, Haaken grew impatient. No closer to controlling Iceland, he eventually ordered Snorri killed. Snorri died, murdered in his own home.

The sea battle, of 25th June 1244, came three years after Snorri's death. His nephew Þórður, based in Westfjords, decided to sail east to Eyjafjörður to reclaim family land, and to exact revenge for the death, in battle, of his brother.

As he sailed across Húnaflói, toward Skagi, with fifteen ships and 210 men, he saw Kolbeinn's fleet approaching. Kolbeinn Arnórsson came from Skagafjörður, the fjord to the east of Skagi. With twenty ships, his men outnumbered Þórður's by more than two to one. The two fleets engaged in the only sea battle in Icelandic history. It is known as Flóabardagi, *the Gulf War*.

The fighting, which included throwing rocks from the boats, continued for more than five hours until Þórður retreated. All the men exhausted, Kolbeinn's ships let them go. Both Þórður and Kolbeinn survived the battle. Although there was no clear victory either way, Kolbeinn's side suffered the most casualties, with eighty dead.

I wondered how large the ships were. On average they carried fifteen to twenty men apiece, so if any were small, like the ones we saw at the farm at Gjögur, the others must have been much larger. I also wondered, who would load rocks as weapons in case of an unexpected sea battle? More likely the larger ships carried rocks as ballast, and the fighters used everything they had available as a weapon.

When Kolbeinn died the following year, his son, Brandur Kolbeinsson, took over as clan leader. Þórður confronted him a year later at the 1246 Battle of Haugsnes, the bloodiest conflict ever fought in Iceland, when about one hundred men died. Þórður survived that battle, becoming the most powerful man in Iceland. As such, King Haaken summoned him to Norway to bring him within his sphere of influence.

The fighting did not subside until 1262, when the Icelandic chieftains gathered at the Althing. There they signed an agreement, the *Gissurarsáttmáli,* accepting the sovereignty of Norway. [4]

As we continued, making visually imperceptible progress from hour to hour across this thirty-mile gap, I longed for each two-hourly snack, and welcomed every opportunity to pump my cockpit dry. Meanwhile, I imagined those troubled times. Somewhere out here, that sea battle had taken place with arrows, spears, axes, and swords. Men had hurled boulders to break boats, shields, and bones.

After eight hours or more of hard paddling, thoroughly chilled, we landed to stretch, and snack, but feeling even colder on land hurried back afloat. Pressing on for another ten miles around the end of the broad peninsula, Skagaheiði, *heath peninsula*, to its eastern side, we were more than ready to quit for the day. Shortly after passing the lighthouse at Skagatá, *toe peninsula*, we spotted a beach. A single a house stood at one end. It was a relief to turn toward shore.

With groans, we ran aground at a respectful distance from the house. In twilight, bitterly cold, I very much looked forward to the comfort of my sleeping bag. To my dismay, the whole beach lay buried beneath deep banks of stinking seaweed. The recent rough seas must have piled them up.

Faced with the alternative; launching again and finding somewhere else to land, we waded, sludge-squelching and skidding through the wet slimy morass, carrying our kayaks to a shingle berm. On the flat top we cleared a patch for the tent, as well as we could, scooping aside as much of the seaweed as possible. We could not clear it all, for the wet fetid fronds clung

Iceland by Kayak

tenaciously to the stones, and more lay trapped between and buried underneath. Having done our best, we pitched the tent, whereupon the remaining slime-covered rocks filled the tent with stench. Craving a brew, we realized we had carried too little water with us. Not finding a stream, and the night being too gloomy to see the state of the water in the lake beyond the beach, we skipped cooking and went to sleep.

In the morning I woke to the sound of rain, but when I opened my eyes, I saw the tent full of flies, their pinging against the fabric maintaining a steady rattle. There was a putrid stink of seaweed. When I rolled onto my side, I could see the gooey paste of squashed seaweed around the groundsheet. It crawled with flies and writhed with maggots, the larvae of these seaweed flies. I felt nauseous. I was quick to get out of there, only to find the seaweed covered beach also swarmed with flies.

Geoff, after knocking at the house with no response, filled a bottle with water from the lake. This water teemed with tiny creatures swimming around in it, so when we saw an elderly man approach the house, dressed in his Sunday clothes, Geoff hurried over to ask for water. The man brought out a red plastic bucket and walked to a stream, quite hidden in the grass near his house. Geoff tipped away the lake water in favor of that.

Nevertheless, we boiled the water to sterilize it, running the stove. Consequently, our fuel ran low after breakfast, and we did not steam our suet pudding for long enough. When we came to eat our lunch later, we realized in dismay that the pudding had barely cooked even on the outside. The gooey dough inside was quite inedible.

This far on our trip, according to the 1761 Homann Heirs Map of Iceland showing the administrative quarters, we had crossed from the West Farthing into the North Farthing, from the Westfjords to the north coast. Having completed what promised

Siglufjörður

to be the longest crossing of the north coast, I felt confident to tackle Skagafjörður next.

We slid silently from the soft oozing seaweed bank onto the water, where the fresh breeze blowing out from Skagafjörður threatened to push us north, out to sea. With no weather forecast, for safety, instead of crossing right away, we paddled south along the coast for an hour or more, to Ketubjörg. There, falling from the cliffs, a waterfall hung like a curtain pushed to forty-five degrees by the wind. Since the wind was invisible, I imagined gravity had shifted a little today.

Sensing we had paddled far enough south to allow for wind drift; we turned, aiming across the fjord toward the distant mountains. Our angle, ferry-gliding across the wind and waves, made for easy paddling but took us too far south. When we compensated, readjusting toward Málmey, *metal island*, the seas hit us from the side; awkward, and uncomfortable. Soaked for four hours, both of us became cranky, frustrated by weather and sea conditions over which we had no control.

Not until we drew Málmey abreast to the south did the wind ease. The sea aligned more from behind us, our kayaks began to run freely on the waves and our tension dissolved. According to the Admiralty Pilot, this area has magnetic abnormalities that make the magnetic compass appear dead. The island's name, Málmey, *metal island,* suggests why.

We passed a lighthouse at Straumnes, *stream nose or race point,* and landed in a little pebble bay to snack on cheese and inedible dough. There, the wind blew relentlessly through my wet fleece leggings, and even Geoff, in wetsuit trousers, grew impatient to relaunch. We cut across the next bay, about ten miles, toward a block of dark, jagged, snow-covered mountains. Glowering shower clouds bubbled, and hovered, dragging gauzy curtains across the peaks. The mountains behind the bay looked rusty brown, much more rounded, with steep diagonals of scree, and grass. As we closed, the far shore began to look greener than was apparent from a distance, with spiky pinnacles guarding the slopes. The cliffs finally began to offer respite from the biting wind. Switched off, paddling automatically, we approached under the red sector of the lighthouse at Sauðanes to creep around the convex coast into the mouth of the fjord, Siglufjörður, weaving between rocks and sea stacks.

"Kippers! I can smell kippers!" Was it the smell of fish drying on the racks, or odors from a fish processing plant, wafting out? Whichever, it was tantalizing. "I'm starved! I could die for a pair of golden kippers, with butter, and thick slices of fresh crusty toast!" When Geoff did not respond, I just felt hungry. The tightness clenched my stomach all the last few miles toward the town, Siglufjörður. The uncooked dough had been indigestible. Aware of the reason it was undercooked, I knew we could look forward to no more than a brew tonight with what little remained of our fuel.

"We'll have the best chance of finding fuel if we paddle into the harbor," Geoff suggested, so we rounded the harbor arm. Dazzling lights shone especially from over the fish factory, which gushed out a fish-scented, steamy white cloud. Fishing boats passed, entering, and leaving. There we ploughed into an oily slick that smelled strongly of fish.

"Ugh! This is revolting," I moaned as my paddle slipped, greasy in my palms. We carefully avoided the fishing lines draped from the back of a vessel at the quay, passing a trawler from Reykjavík and another from Grindavík. Beyond stretched long wooden docks and derelict jetties with warehouses and sheds, all of timber. The few jetties that had not partially collapsed into the fjord looked as if they were ready to follow the trend.

Beyond the extensive wooden docks, and looking unreal in the half-light, was a sturdy jetty with a couple of large steel-hulled stern trawlers from Siglufjörður, painted orange and white.

Siglufjörður

Seeing the harbor busy, and today being a Sunday, I recalled watching the fishing fleet leave from the Scottish port of Peterhead in Scotland on the stroke of midnight, the moment Sunday was over. Here, Monday, and the start of August, was just hours away.

We drifted to a standstill beside another wooden quay, just beyond the trawlers, relaxing on the gently undulating surface, an oily sheen dulling the reflections. "What on earth are we doing in here? We're right in the center of town."

"Let's just land and have a look."

We selected a wooden quay and climbed, gingerly, to the top. The whole quay was rotting. There were large holes with grass growing between the planks. To stay here, we must haul up the laden kayaks, so we debated how best to do that.

"This can't have been used for quite a while," Geoff mused, once our kayaks were secure. "I wonder if the holes are where people have fallen through?"

"That's reassuring." I stepped forward gingerly, testing the firmness, feeling the planks give a little under my weight.

We chose a more solid part of the quay to set up; a place where grass grew on the wood, finding bricks to pin the tent. Everything around was soaking wet.

"They've had quite a bit of rain here, by the look of it."

"Yeah, we've been lucky today. All those showers crossing the mountains, and hardly anything on us."

We laid a kayak along either side of the tent, creating the illusion of security for the entrance. When a group of teenagers came to see what we were doing, we asked where we could buy *bensín*, Icelandic gasoline, for our stove. They did not know of anywhere open.

"You could ask at the police station. I'll take you there," one offered happily.

It was not far, but he stopped short. "You go in," he encouraged, holding back. "I'll stay out here. I've been drinking, and I'm too young."

Inside, we were disappointed. "Bensín? No, nothing will be open at this time of night." He paused before continuing, "We do have some here you can have." He poured us two liters, in the back room, refusing payment. We carried it away, elated. We could cook and eat.

By now it was late. Weary, having paddled more than forty miles, and replete with hot pasta and canned fish, I stretched out comfortably in my sleeping bag on the luxuriously flat surface. But an uninterrupted sleep here in town? A church bell woke me every hour throughout the night. The chimes sounded so clear and steady, it was no surprise to greet a gentle morning, with little breeze, and low clouds. Fetching water from a dock hose nearby, we made breakfast, surrounded by children who inspected the kayaks and nosed into the tent.

Strolling into town to shop, we learned that today was a public holiday, everywhere likely closed. The police officer on duty suggested one possibility and locked up the police station to take us there himself. To our disappointment, this store carried almost no food. When the shopkeeper learned of our kayak trip, he phoned the owner of a nearby store, who came to open it for us. In the end, we were able to buy all we needed.

On the way back, the officer explained more about the town, and how it grew with the herring fishery. "Every fishing season, the whole area around the old docks, where you are camping, used to look like a forest. There were so many masts, with boats from everywhere. Everything grew fast with the herring. In the thirties and forties, the heyday, the town spread to about three thousand people. There were more in the summer of course, at least ten thousand, what with all the workers.

Siglufjörður

"There were twenty-three salting stations, and five factories to process herring into meal and oil. Then, in the fifties, the catches dropped away, and it all seemed finished. People started moving away.

"In the sixties, the herring came back, and we got going again. With all the new fishing techniques, the catches were huge. There was a peak in sixty-eight, an incredible season, but the next year? Nothing. No herring. None. More people left town right away, and people are still leaving. About one-third of the town is empty."

He stopped and pointed. "Down there, you used to see all the masts. The herring ships came in, day and night, so the shops stayed open twenty-four hours a day. There were two hotels, one large, one small. Everyone in town did well. There's only the small hotel now, the big one closed. Those warehouses down there, they're empty, falling apart. People hang fish there to dry."

He seemed proud to describe the scene of doom and gloom. "Yes, people left, the foreign ships left. We were a shanty town. It is only in the last few years that cod has brought things back a little. You know, a little." He turned, lifting his hand in front of his face, and holding his forefinger and thumb barely apart. "Just enough to offer hope. Enough to make you think things might pick up a bit."

"There are an awful lot of houses boarded up," I agreed.

"Yes. We are down to about two thousand people living here year-round. We've got empty houses, everywhere."

This adjusted my perspective on the cod wars. Could Iceland, by no longer competing against foreign vessels, manage her own fishing fleets without wiping out the fish? Was it possible to develop the cod fishery enough to sustain these former herring towns? If the herring ever returned, could a fishery be sustained under more control?

At the tent, I mixed a pudding of flour, suet, and raisins in a small pan, and set the pan in water inside a larger pot to steam.

Iceland by Kayak

Packed and changed, but loath to leave the stove burning unattended, we took turns to explore. On a drying rack in one of the derelict buildings, a row of cormorants hung alongside the fish. Had they become trapped in fishing nets, or caught on lines like those flung from the trawlers docked against the quay?

Our pudding cooked, we floated the kayaks and climbed in. A crowd gathered to watch us leave. People of all ages watched quietly, helping when they could, passing down a paddle, steadying a kayak. When we drifted into the harbor, waved, and called goodbye, a quiet chorus of "Bless," *goodbye*, breathed back to us like the break of a gentle wave.

We passed the derelict quays where the slowly rotting forest of trees, which shored up the platforms from underneath, smelled of decay and of long-dead fish. Where posts had splintered or leaned, the dock above sagged, or opened into holes. Loose timbers dangled into the water, groaning, swayed by the swell. Neglected, these docks were in danger of collapse. Then, what would happen? Would the timbers float away, or waterlogged, sink?

Before leaving the harbor, we paused by the final and well-maintained quay. Fat pipes pumped fish from a large trawler up to conveyer belts, which hurried the fish into the adjacent factory. There was the clamor of moving machinery, the shrieks of gulls and a steady hum from the ship. This was a world and a time away from the wooden barrels and fish boxes of the small towns. Here was a new direction for Siglufjörður.

Siglufjörður

BOATS PREPARING TO LEAVE FLATEY ISLAND.

HÓLMGEIR'S SOCKS. (PAGE 298).

MAP 9. ICELAND, SIGLUFJÖRÐUR TO LANGANES.

26
Flatey Island

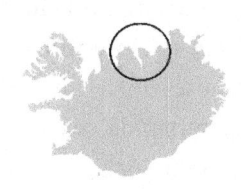

Low clouds still covered Siglufjörður when we reached sunshine at the end of the fjord at Hella, to begin a slow turn toward the east. We had left early against local advice to wait until five or even seven in the evening. That advice was to help us avoid a headwind, which usually sprang up from the east in the afternoon and dropped later. Knowing how cold paddling could be at night, we chose to leave early anyway.

Now, luck was with us. As we prepared to cross the mouth of Eyjafjörður toward Gjögurtá, the northernmost point of land on the other side, the wind blew from the northwest.

Eyjafjörður means *island fjord*, named on account of Hrísey, the second largest island off the coast of Iceland. Hrísey stood ten miles south from us down the fjord. Twenty-five miles beyond Hrísey, at the southern end of this, the longest fjord in Iceland, spread Akureyri, Iceland's northern capital. The first settler at Akureyri was a man from Ireland, known as Helgi the Lean, in the ninth century. Since Helgi Eyvindarson is a Scandinavian name, he might not have been Irish, even if he arrived from there.

Akureyri is petite compared to Iceland's other major cities, which all lie in the southwest. Reykjavík, Hafnarfjörður and

Iceland by Kayak

Kópavogur, together boasted a population of more than 100,000. That is close to half Iceland's total which was 218,000 in 1975. Those three close neighbors, if coalesced into a single conurbation, would make Akureyri, with its population of just 10,000, Iceland's second largest city, rather than its fourth.

The wind blew from the most awkward quarter behind us, making it difficult to steer a straight course. Each time my kayak took off down-wave, it skewed toward the wind, and I had to force it back on track. This pattern: let fly and rein back, held my attention until I spotted a dolphin approaching.

My eye on it, I quickly realized it was part of a pod. In time the dolphins surfaced close to our kayaks. Had they spotted us and come close, investigating out of curiosity, or had our paths randomly crossed? From how far away could they see our kayaks? Were we more visible to them, or vice versa?

The dolphins distracted us until completing the crossing we stopped on a ledge that extended from an isolated sea stack. The moment our kayaks were secure, Geoff turned to me with a broad smile, to ask eagerly, "Okay! Now, where's that pudding?" I pulled it from my deck bag, and he dug in hungrily. "Oh, this stuff's much better! It's cooked," he said. "That last lot was disgusting."

There was a hole through the stack, so we climbed up to peer through it at the murky headland beyond. The narrower view helped me spot details I might have missed when taking in the wider panorama. But high up on the stack the wind cut like ice, discouraging me to linger. "Let's get going again, Geoff," I pleaded. "It's too cold to hang around."

I slid my kayak across a cushion of seaweed onto the water. Kneeling on the seat, I corkscrewed into the damp cockpit and sealed myself in. Rinsing the seaweed slime from my hands, in finger-numbing water, I paddled vigorously to warm up, aiming for the gloomy cliffs.

Flatey Island

Gradually, the wind gathered strength, pushing at our backs yet still chilling. We forged on, watching a shadowy shape materialize in the distance ahead. Flatey island, we realized, lay within ten miles. Flatey was an obvious target. "It'd be nice to camp on an island," Geoff had expressed earlier. I love islands too, but I was uncomfortable. Surfing on waves which heaved and lurched ahead of the freshening northwest wind, I grew ever colder.

I accelerated, took off on a wave and steered; a hissing plume chasing my trailing blade. One wave intersected another like a pair of scissors closing on me. Where the edges met, the wave jumped and steepened. For a pulsing moment, it shouldered my kayak and pitched me faster downhill. As I sped onward, chasing the deepest hole, spray flew from my bow, showering me. Water jetted up my sleeve as I steered.

Soon enough my kayak stalled, but as one wave crept ahead, another followed right behind. The kayak still carrying momentum only needed a couple of swift paddle strokes to send it on its way again, flinging more spray into the air.

When I ran out of drive at the end of that set, I looked back. Geoff was far behind. I took the opportunity to lift my elbow and drop my hand, stretching open my elastic cuff with numb fingers to pour the water out, from one sleeve at a time. This was the best routine, pouring it straight back into the ocean. Otherwise, each time I lifted an elbow to shoulder height when steering, water ran in the other direction. Like an electric shock to my armpit, it ran down inside my fleece. Seeping onward, it gathered in my cockpit, sloshing ever deeper.

Lost in thought, I heard a whoop from behind and turned to see Geoff speed past on a wave. Before I caught a suitable wave to get going again, Geoff was far ahead.

Iceland by Kayak

It was getting darker and colder, the clouds lower, the weather ominous. Coming within hailing distance of Geoff, I shouted, "My hands are numb, numb, numb!"

"Mine too," he called back, agreeing, "It's a wee bit parky tonight, isn't it?"

We reached Flatey at two in the morning, slipping into the calm of the harbor at the southeastern end, guided by two beacons. There were boats tucked in there, one a largish wooden line-fishing boat with a small open boat tied at its stern. Another was a whaler with a harpoon gun mounted on deck at the bow. While curious to look more closely at the whaler, I was too preoccupied to investigate right away.

We pulled the kayaks ashore and dragged them across the grass. The island was low and flat, with nothing to block the wind between Greenland and us, except for a small engine house just yards along a track. The wind cut into me as if I were naked.

"I'm going over there!" I pointed, my body shaking. I dragged out dry clothes and, leaving everything else scattered beside the kayak, hurried to gain what shelter I could find, behind the small building. I changed, awkwardly, with numb fingers.

"This is painful," I shivered, as we pitched the tent between the kayaks. I carried boulders to hold it down in case the wind strengthened. I did not want the tent, so exposed, to end up in the harbor. Ducking inside, I climbed into my sleeping bag before lighting the stove in the tent entrance to get a meal started.

Late, and cold, it was not the time to explore. "I'm going to take a closer look at that whaler in the morning," I resolved. I was to be disappointed. When the sound of a boat engine awoke me in the night, I ignored it and missed my chance.

Awakened later by the sound of voices and curious to see, I unzipped the tent. A man and woman, with children, had gathered by Geoff's kayak. I called good morning, and they came closer. Since it was raining and windy, once Geoff sat up, I invited them

Flatey Island

all to shelter in the tent. It was tight, but everyone managed to squeeze in.

"We are on Flatey, staying at our holiday house," they explained, their wet clothes steaming in the cramped space. "There's a government restriction on commercial fishing for cod, so we take a break."

"Where do you usually live?"

"Húsavík. Most of the islanders from Flatey moved to there. There's just one farmer who keeps animals here; he stays for three weeks each year. But everyone's left the island. The fish oil plant closed, and the school's empty."

"Do you know the weather forecast?" I asked.

"Ah, it's not so good. Winds from the northwest, force three to five, with rain, and it'll be cold."

"That sounds uncomfortable," I said. "We might stay here today."

After our visitors left, we sat in the tent entrance, which was drier than the back end where the wind blew water through the fabric. When cooking breakfast, the sound of voices alerted us again. A slender man, cleanshaven, with short fair hair, blue eyes, and a narrow face, introduced himself as Guðmundur. He was with his young daughter. "That's my fishing boat there," he pointed proudly at the wooden fishing boat, the *Aron*, moored at the quay. He crouched into the tent with us, where we described our journey so far on the map. When I asked about his boat, he invited us to follow him on board.

The well-kept *Aron* was sparkling clean. The tiny wheelhouse held the radar, echo-sounder, a half-sized chart table, and the controls. Below was the immaculate engine room, the powerhouse that could propel the vessel at ten knots under full driving force.

A separate companionway led down to Guðmundur's cabin, which was compact and basic. Two holes, curtained off behind the bench seats either side, each allowed access to a bunk compartment with neat racks inside for books and small items. The cabin itself had a couple of built-in cupboards, a sink, and a table.

"My ten-year-old son uses the second bunk in the summer months when there is no school," he explained.

The crews' quarters were in the bow, six holes accessing the bunks, two long wall-benches, and a table. At the wide end of the triangular space was a large cooker, with a huge stainless-steel pot held in place by metal rails.

"One man is the cook, nothing else. The other men take turns on deck, using the hydraulic reels. Each line has seven cod hooks, and a three-kilo lead weight on the end."

Between the crews' quarters and the wheelhouse was the fish hold. The *Aron* loads up with ice to freeze the fish, and leaves Húsavík for three or four days at a time, sometimes for a whole week, coming back when the ice begins to melt, which is usually before the hold is full.

When Guðmundur divulged that he fished mostly on Grímseyjargrunn, between Grímsey and the mainland, or around Langanes, I asked, "How do the currents run around Langanes?"

He traced with his finger as he explained. "The Gulf Stream, here the Irminger Current, flows clockwise around Iceland. It goes west along the south coast, north up the west coast, east along the north coast, and south down the east coast. The East Greenland Current joins the Irminger Current here," he pointed, "in the northwest, flowing east, and the East Iceland Current meets it at Langanes, flowing south. So, at Langanes, all the currents run together past the point as a race. The flow is always out to sea, but the direction depends on the tide. The safest, the most comfortable place to be, is right under the cliff where there is only a narrow section of rough water."

"That's useful to know," I glanced at Geoff. "Not that we're likely to stray far from land around a headland like that."

Flatey Island

Back on deck, Guðmundur stopped to check the fish that hung by their tails from the rail in the rain. He unhooked four, stuffed them into a plastic carrier bag and handed them to us. "They'll last for about a week," he explained. "Boil them for about ten minutes. We hang them for two or three days like this to make the flesh firmer, with a different flavor. At sea, it makes a change from eating fresh cod all the time."

"Come, you can see my storeroom." He walked ahead through the wet grass for two hundred yards to a line of small buildings along the shore of the western harbor arm. Inside his warehouse stood square stacks of salted fish, and barrels holding more. "Here," he said, lifting a flattened slab of yellowish fish from the white pile, showering coarse salt crystals from it. "This will keep for much longer, dried and salted like this, but you must soak it in fresh water for a couple of days to get rid of the salt. Change the water." His eyes turned to us and lingered fiercely in caution, before he set the fish back on the stack.

He pointed to the racks of fish drying. "Those stockfish, we dry for three months. An Icelandic company buys them from us and sells them on to Nigeria, and Italy. They are the smaller fish, the bad fish. Icelanders never eat them."

Guðmundur walked us back to our tent, before on impulse inviting us to go with him to his house for coffee. "We can listen for an up-to-date weather forecast on the radio," he suggested, checking his watch for the time. As we walked, he pointed out the hand-pump the islanders used to get fresh water from the well.

Leaving our shoes at the door on entering the house, the usual courtesy, I felt ashamed of my wet socks. My shoes, falling apart, held together with string to keep the soles on. My socks could not get wetter. Should I remove them too? That would reveal my white, prune-wrinkled feet, covered no doubt in stray strands of wool. I kept my socks on.

Iceland by Kayak

Guðmundur's happy family clamored to greet us. Guðmundur introduced his father, brother and sister, his wife, and his siblings' spouses, plus all their children. They hung our wet anoraks and handed us towels to dry our faces and hair.

"Sit down and get warm!" they encouraged, before presenting their guest book for us to sign.

Over coffee, everyone fell silent to hear the weather forecast on the radio. "It's going to get worse," said Guðmundur, summarizing the essentials in those few words.

Guðmundur's father, Hólmgeir, who until now had sat beside the stove watching, and listening, raised himself stiffly from his chair and approached. He bent down to feel my feet. Without a word he crept back to his chair where, bending down with a little difficulty, he pulled off his socks. These he offered to me, encouraging me in Icelandic, and with gestures, to replace mine with his. He took my wet socks, hung them by the stove, and sat back down, barefoot. I felt deeply moved.

"Let us show you the house!" someone said eagerly. We followed behind a whirlwind of excitement as the family traipsed from room to room, chipping in extra snippets of information at every opportunity. The children were especially delighted, even though not all spoke English.

Hólmgeir built the house when Guðmundur was born. It still had the original wood paneling in two rooms. The family moved out years ago, but on return visits began rebuilding. If all goes to plan, they hope to finish the job in two or three more years. So far, they had replaced rotten timbers, and lifted one side of the roof to make space for three rooms upstairs.

When I admired the crocheted snowflake doilies decorating the upstairs windows, Guðmundur's wife smiled and said she had made them. "Look, I'll show you how." She took out a partly finished piece, and a crochet hook.

Flatey Island

Mesmerized, unable to follow the magic of the hook, I confided, "My grandmother crochets too," recalling her wonderful shawls and cloths that seemed to grow from her fingers.

As we sat by the window, the cloud began to lift from the mountains on the mainland, revealing a fresh dusting of snow.

"We're going to have to leave today," Guðmundur announced, standing up abruptly. "We should get back to Húsavík before the weather closes in." At this announcement everyone began packing. Setting aside food that would spoil if left, they gave this generous supply to us. Guðmundur's father handed me back my dried socks, but when I bent to take off his, he stopped me. I must keep them.

My own looked crisp and stiff. When I rubbed them together saying, "Look, English wool," everyone laughed, including the eldest girl who, caught by surprise while eating, spluttered out a mouthful of tomato sandwich. She blushed in embarrassment.

Geoff and I helped carry cases of luggage to the *Aron*, stopping briefly at the new stone jetty while Guðmundur looked across at a boat there. Previously his own, it carried the same name as his current vessel, *Aron*. Her crew stood at the bulwark, gutting fish. As the slender blades flashed, the cod heads fell heavily, and the guts flew out onto the water. A flock of fulmars wheeled and dived, beak first, packing into a tight struggling knot to seize each morsel before it sank.

All the Flatey inhabitants had moved to the mainland by 1967, shortly after the completion of this new jetty. The structure has proved valuable since, in rough weather. Boats often come here to gut their catch in shelter instead of suffering out on the fishing grounds.

Guðmundur's brother invited, "Would you like to use my workshop, to sleep and eat inside instead of your tent?" In view of the weather, we especially appreciated his kind offer. Apart from the extra space, we would welcome somewhere drier than our leaky tent. He opened the door to a room full of wooden barrels and handed me the padlock to snap shut when we left.

He sped off in his small boat, followed by more of the family in the small red, and white, wooden fishing boat, *Kristinn*. Not until the tide lifted *Aron's* keel from the bottom did Guðmundur leave, with the rest of the family, towing Hólmgeir's open boat. They waved until they were well under way.

"What a wonderful family," sighed Geoff. "They have such great energy!"

Now we were alone. It was Tuesday. We could hardly imagine that Saturday would come before the weather cleared us to leave.

We fetched our sleeping bags, mats, and stove, and cooked a chicken stew in the workshop, from ingredients the family had left. Cleaning the stove afterward, I donned an extra sweater to try to get warm. When that was not enough, I sat in my sleeping bag.

The weather closed in as forecast, blowing a gale from the northwest with driving sleet and rain. The warehouse afforded shelter from the wind, but not from the cold. We waited, bundled in sweaters, huddled in our sleeping bags, chatting, writing in our journals, and making hot drinks.

Stepping outside became a breathtaking excursion. The wind, funneling past the building, almost took us from our feet, and rain drenched us. Hastening back, invigorated, we would invariably heat another brew to get warm.

To refill our water bottles, we took turns to sprint to the pump. I could not get water to flow, so I lifted the wooden hatch, beside it, to the well shaft. With no bucket to lower, I climbed inside to a narrow concrete ledge, and down the stonework to the water. Crouched there, braced just above the water, I filled the containers. If uncomfortable, trying not to fall in, it was noticeably warmer down there, out of the wind.

Curious to see the ocean, we decided to visit the lighthouse for an unobstructed view to the east. The only signs of life on the

Flatey Island

island were the small Icelandic horses, and sheep, although the houses, boarded up against the weather, looked quite new. Most had a metal water butt outside to collect rainwater from the roof, a smart alternative to walking to the pump. Peering through the windows of the little school, we could see books and papers, tiny chairs, and desks, waiting for the missing children as if left only for the weekend.

Crossing the island, I felt like a ship at sea, the deep grass blown into wild thrashing green waves around me. But it was a calm sea compared to what we saw when we reached the lighthouse. Amid the thunder of waves, the ocean, heavily furrowed, dark, and forbidding, trundled past the island. Off to the north, the air billowed with flying spray.

Already wet by the time we returned, we stayed outside a little longer, pulling our paddling clothes from our kayaks to hang from an old trailer outside the warehouse, to rinse in the rain. We also positioned containers to collect rainwater, so we would not have to visit the well so often. Back in shelter, I began to unhem the cuffs of my flared jeans, cutting spare fabric to stitch around my shoes to hold them together.

"Geoff, did you see the double-glazed windows here? I only just noticed."

"No," Geoff replied, "but it doesn't do us much good does it? It's freezing in here. Surely, the weather can't keep this up much longer, can it?" He fiddled with the stove thoughtfully. "Why don't you mix a suet pudding in case we can leave tomorrow? It'll be an excuse to run the stove for a bit."

He perked up, "You know, pudding will be okay, but what I really fancy is a huge cheesecake."

"Would you like that with fruit?"

He thought for a moment. "Blackcurrant this time please. And with cream."

Iceland by Kayak

The rain stopped late the following afternoon. We rolled up the legs of our jeans and waded through the wet grass to the quay. There we learned from a fisherman, taking shelter on his boat, that the weather should improve soon. Reaching the locked church, and peering through the window, we saw an impressive congregation of gleaming flies, dead, scattered across the floor.

The spluttering of a boat engine signaled the departure of the fishing boat, leaving us alone again. Not long afterwards the wind died, the sky cleared, and the sun warmed the evening. We flung open the workshop door and strung a line to hang our gear in the sun. Across the sound, the mountains shone with snow.

Perilously close to running out of fuel, I remembered seeing an old Toyota Landcruiser. There it stood, abandoned, the matt color of faded red-lead paint, with a white cab, grey-green roof, and rusted luggage rack. The windshield was flat glass, starred and cracked. A length of frayed rope, faded blue, lay tied to the front as a towline. I checked the fuel tank and smelled gasoline. Searching around, I found and cleaned a length of hosepipe, and a bucket, and siphoned out a couple of liters of fuel. We could cook freely again.

Restless, holed up for too long, we explored. Geoff spotted an old red McCormick tractor and climbed up onto the metal seat. It had no provision for a cab. He looked down at me with a look of contentment. "I'd love to buy one like this, repair it, and drive it home," he confided. He sat there, comfortably relaxed, with the steering wheel in his hands. He looked ready to start on his way. I could see him, bouncing along a narrow country lane somewhere in rural Britain, puttering at fifteen miles per hour, indifferent to a tailback of motorists impatient to pass.

I left him daydreaming on the tractor and explored the structures along the shore. One looked to be the remains of the fish oil plant. All the buildings held fishing tackle. Fish hung to dry too, alongside wooden barrels, and neat stacks of split salted fish.

Flatey Island

The waist-high stacks; blocks coated in glittering salt crystals, looked frozen.

The evening was glorious. The sun dried our paddling gear in no time, so I washed and hung other clothes. Then, I joined Geoff, wading through the grass across the island again, past the empty houses, turf-roofed buildings, and the school. The grass gleamed in the sun. Horses with glossy coats chased each other, kicking and bucking, twisting, and turning in play. Everywhere I saw sheep. It was uplifting but at the back of my mind, I wondered about tomorrow. The weather, wonderful at last, could be so fickle.

AFTER THE STORM, FLATEY.

AUGUST SNOW SEEN FROM FLATEY HARBOR.

AUTHOR ON FLATEY ISLAND. .

27

Fox Plains

Waking in the tent after our final night on Flatey, I delighted in clean, salt-free kayaking clothing. But not for long. Kneeling on the seat of my floating kayak, corkscrewing precariously into the tiny cockpit, I overbalanced and faceplanted into the harbor. Geoff laughed so heartily; I too saw the funny side.

Leaving Flatey Island, we slid gently onto the open water of Skjálfandi, *shaking bay*, a known earthquake-prone zone. The weather had a crisp autumnal edge, and the sun dazzled back from the snow-covered mountains beside us. Across Skjálfandi, the distant shoreline drew a low shadow. We drifted, taking it all in. For once, the dark waters undulated, the surface silky with swell, not wind-chopped, or percussive.

I could see where the bay narrowed slightly to end at a seven-miles-wide, east-west beach, the edge of a low plain reaching far south. The snow-clad mountains beside us ran south, like a rib, bordering the plain on this side. Alongside ran the fourth longest river in Iceland, the Skjálfandafljót River, *trembling river*, which begins as meltwater from Vatnajökull and empties into the corner of the bay.

Another mountain ridge bordered the eastern side of the plain, with the Laxa, *salmon river*, running alongside. The Laxa would meet the ocean at the far corner, just four miles from where Húsavík hid from our view tucked into a small bay near the base of the Tjörnes peninsula. Low cliffs defined much of the distant shoreline of Tjörnes.

We both wanted to visit Húsavík, one of the first places settled in Iceland. Guðmundur's family, from Flatey, lived there. But today was already the sixth of August, time slipping away. Instead of thirty-one miles around the bay, with an overnight stop, we chose to cut straight across at half the distance.

To keep us going we carried our suet pudding, cooked with sultanas and orange peel. To break the crossing into measured segments, we would stop every two hours for a sip of water and a chunk of pudding. Having agreed a target, and a compass bearing, we set off toward Tjörnes.

The kayaks gliding smoothly, allowing for easy conversation, we began debating paddling techniques. We did not always agree. For example, when reversing a kayak, I posited that, since the face of a paddle blade grips the water well, and the back grips poorly, we should use the face when reversing.

"If you use the back, it'll slip in the water and be less effective."

"No," insisted Geoff. "It's wrong paddling technique because you have to change your hand grip."

"Why does that matter? You can change back when you want to paddle forward again can't you?"

"What happens when you react suddenly? You'll find your blade the wrong way around. You must keep the same grip, to know how your blade is, whatever you're doing."

Fox Plains

"Oh, come on! How difficult is it to turn a paddle around when you need to? And even if it is tricky sometimes, that doesn't make it wrong. You shouldn't avoid strokes that are harder to learn. And what about a Pawlata roll?" That technique, named after the Austrian kayaker Edi Hans Pawlata, gains extra leverage from an extended paddle, moving the hands toward one end. "Or the *put-across* roll," another extended paddle roll. "Do you think they are wrong too, just because you change your grip?"

"Isn't it better when you don't have to change your hand grip to roll?" he countered with a smile, knowing I would have to agree.

"That's more extreme. Reversing, you only need to turn the shaft in your hand, you don't have to move your hands along to the end." But Geoff had cracked my argument.

The conditions allowed us to continue chatting until near the far shore, where a sea stack of tall, straight pillars of grey columnar basalt attracted us. Circling it, we saw a perfect ledge at the base on which to land.

The top of each basalt column formed an almost flat, slightly dished platform. It was easy to scale the stack, scrambling up in steps of never more than five feet. From the top we looked down at the yellow decks of our kayaks parked across the honeycomb pattern of the wave-cut shelf. From up high, the columns we climbed looked like a giant drum kit.

The dark rock had absorbed the heat of the sun, so we perched comfortably, dangling our legs over the edge, nibbling at our food, and studying our maps.

About a mile into the bay from where we sat stood Hallbjarnarstaðir, the farm below a bluff known for its fossil beds, and likely where the fossils we received at Kollsvík came from. The fossil-bearing strata lie sandwiched between layers of lava. It is one of the few places in Iceland with fossils, mostly marine mollusks. The species recorded at one level show the first

Iceland by Kayak

appearance of Pacific shellfish from the time when the Bering Strait opened between the Pacific and Arctic Oceans. The land bridge separated first around the end of the Miocene epoch, before the start of the Pliocene, 5.2 million years ago.

Iceland straddles the North Atlantic Ridge. Just here, the more recent layers of lava, and the underlying Pliocene fossils, are all part of the Mid-Atlantic Ridge. The oldest regions of Iceland: the east coast fjords on the Eurasian tectonic plate, and the northwest fjords on the North American plate, drift apart at a rate of one inch per year. As the joint tears, magma rises. Lava fills the cracks, cooling into new rock. Earthquakes and volcanic activity in Iceland predominate along this line crossing the country from Vestmannaeyjar and Reykjanes in the southwest, to the coast between where we sat and the next promontory, Melrakkaslétta.

Running south from Iceland, that same seam: the divergent tectonic plate boundary, rips and rewelds itself along the ocean floor along a path midway between continents, all the way to just north of Antarctica. The line extends northward from Iceland too, through Grímsey Island, to the fast-eroding rock Kolbeinsey, and beyond. Iceland, like some humans of a certain age, is spreading from the middle. The next new lava flows on the north coast are most likely to appear right around here.

Seabirds clamored excitedly. From below came the calming echo, like slow sleepy breathing, of waves licking the base of the rock. All the sounds gathered, focused and clear, into the hollow below us. The air carried up the odors of warm rock, caustic bird lime, and drying seaweed.

We sat back, lounging close to the top, for more than an hour before scrambling down. Comfortably warm for a change, we slid our kayaks from the rock, unencumbered by anoraks.

Close to Tjörnes lighthouse, we discovered caves and an extraordinarily wide natural rock archway. With such a long span, I wondered why it had not already collapsed in the middle, although that did not deter me from lingering beneath it. Preferring to explore rather than make miles, we probed in and out of the caves until we heard voices calling, in English.

When we emerged from under the cliff, to look up, we saw people peering over the edge at us, Icelandic hikers, enjoying the weekend weather. They had read about us in the newspaper, so when they spotted us from the clifftop, they hailed us in English. They asked where we had come from today, and where we hoped to reach. Curiosity satisfied, they wished us good luck, and continued their walk.

Up until that unexpected moment of contact, we might have been alone on the planet; so absorbed probing beneath the cliffs. Now, like self-conscious schoolchildren, we redirected our kayaks and focused on paddling around Iceland.

The conditions deteriorated as we began crossing Axarfjörður, so we rafted for balance to don our anoraks. Paralleling the shore, six or seven miles out, we aimed for a mountain, trimmed flat on top by cloud. The weather closed in, whipping up a sharp jittery sea.

Exposed out in the open, paddling became dreary, but we pressed on until, spotting houses at Kópasker in the distance, we stopped to bail our cockpits. The evening felt so bleak that we reconsidered our plan. Despite the miles the detour would add tomorrow, the perceived comfort of the village won our vote. We had covered more than forty miles by the time we neared land at ten o'clock in the evening. I longed to get out of the dousing waves and biting breeze.

Aiming for a sandy beach beside the small harbor arm, we were soon shivering on a patch of windswept grass. There, jumping up and down, swinging our arms vigorously, we wished ourselves warm.

Scouting, to our surprise, we saw a shop nearby and it looked still open. We hastened to buy food, but the girl behind the counter could not understand us, even when we pointed to the items we wanted to buy, displayed on the shelves behind her. Another shopper, taking pity on her, led us to a hotel, reasoning that we should more easily get what we needed there.

At reception, we asked for flour, raisins, cocoa powder, sugar, and onions, to take away with us. The receptionist invited us to sit at a table while she went to fetch. After a long delay, she brought us mugs of steaming hot chocolate, with sliced bread on a plate. Puzzled by the miscommunication, we explained again.

Waiting once more, almost dozing in the luxurious warmth, I breathed in the comforting aroma of hot chocolate. Geoff, yawning, mused wistfully, "What do you think it would cost if we spent the night here? Clean sheets, hot shower...?"

The receptionist told us when she returned with our supplies. A room here overnight would cost us three thousand Króna each, the equivalent of seventy-five international postage stamps. Beyond our budget, the idea was enticing. Reluctantly, we dragged ourselves back to the beach. After the warmth and comfort of the hotel, the bitter wind felt doubly harsh. We hustled in the near-dark to find boulders to hold the tent, which we pitched where it seemed best, beside a gas station. Neither of us wanted to cook anything complicated, so we boiled our usual standby of porridge, with sultanas.

From eight next morning, well wrapped in our jackets, we watched the freshening northwest breeze whip up tightly packed and ever more vigorous whitecaps. "Let's go back to the hotel for a cup of coffee," Geoff suggested. "We can use their loo and fill up our water containers."

Warming, we relaxed, our hands cradling hot mugs. Idly aware of the chatter of voices from the radio above the counter, I suddenly recognized the pattern of the weather forecast and asked for a translation. "Here, northwest winds at fifteen to twenty knots, rain in the evening."

Fox Plains

"Well, that should be okay," said Geoff, breaking into a smile. "If we get going right away, we can stop before it rains."

That did not happen. We saw how the wind had strengthened. The fishing boats in the harbor pitched and heaved, straining at their moorings. Shielding my eyes against the sand and dust, I walked over to the quay to watch a boat unload. The men carefully weighed each crate of fish before hoisting it from deck onto a truck.

The skipper saw us watching and greeted us, so we approached. He gave us his analysis of the sea conditions we might expect. "This wind will be force five to seven on land. At sea it will reach force six to eight. That is not good for us, fishing. We'll stay here till it abates."

Geoff and I looked at each other, discouraged.

"Here," the skipper offered, seeing our disappointment. "Take a fish." He handed us cod.

I thanked him enthusiastically. Fresh fish was so much more enjoyable than our canned alternative. Geoff, thinking ahead asked, "We should move our tent if we stay longer. Is there anywhere more sheltered nearby, somewhere that's okay for us to camp?"

The skipper of another boat, eavesdropping on our conversation, suggested, "There's a sheltered place behind my house, where you're welcome to stay. It'll be better than where you are now. My name's Barði Þórhallsson."

He fetched his Land Rover, while we collapsed our tent, and he carried us, with our camping gear, to his house. Slashed across the land behind his house ran a jagged split in the earth. "We had an earthquake in January last year," he explained. "There was nobody hurt, but it destroyed the building where I stored and smoked meat, and the outbuilding where I kept potatoes. Those houses over there were seriously damaged," he pointed to houses a hundred yards or more away. "On the other side of the far house, some holes opened in the ground; extremely deep."

One broad crack ran within feet of the corner of his own home. I peered into it and realized it was no mere surface crack. Between fringes of overhanging grass, the clean gash in the rock dropped down out of sight into darkness. It had been there for eighteen months. How deep was it, and what was at the bottom? Would he try to fill it, or had he already accepted the new hazard? It made me uneasy.

Barði continued, "I was at sea at the time of the earthquake. I had no idea. You can't feel it on the water."

"You see that low area over there?" He pointed to a sunken area the size of a football pitch. "That was all at the same level as this before. The earthquake measured six point three on the Richter scale."

I looked back at the open cracks in the ground beside his house, where the solid rock had stretched until it burst apart. Which tectonic plate did we stand on here, I wondered, North American, or Eurasian? Or was that yet undecided?

Barði's two little daughters, Þórný and Helga, helped us collect stones to hold our tent, while his wife Anna brought out a flask of coffee. Barði took his leave; his boat still needed unloading. When Geoff and I had finished setting up, unable to relax out of sight of the sea, we wandered back to the water.

It was raining hard by the time we reached the harbor, so Barði invited us on board, where we sat in the cabin, drinking coffee with him while he waited his turn to unload. Our conversation turned to the cod war.

"I worked on one of the gunboats all through those troubles," Barði confessed. "They equip them for rescue work, with reinforced bow and stern, for pushing and towing boats, but that made them suitable for the new work too."

Fox Plains

"We'd often cut the tackle of the British trawlers." He flashed a glance at each of us, gauging our reaction to the sensitive topic, before continuing. "We dragged a short cable, with a four-pronged hook on the end, and we hung a heavy chain under to hold it down. There was a sharp edge inside each hook.

"Those hawsers, pulling the trawl nets, can withstand a load lengthwise, but they snap if bent at a sharp enough angle across an edge. So, we would steer close behind a trawler to pick up the two trawl wires and keep pulling sideways until they snapped."

Geoff cut in, "Isn't that dangerous? Don't the broken ends fly back up? Surely, that could injure people on the trawlers, or kill them?" That was the argument the British media had championed, goading the government into sending more naval frigates to protect the British fishing fleet.

The press contended that a cut trawl threatened more than just the livelihood of the courageous fishing crew. For sure, the cost to replace the essential tackle was considerable, and the time and expense of sailing to Icelandic waters and back, with no catch to bring home, was disastrous enough. But what riled the press most was how the irresponsible actions of the Icelandic gunboats put British fishermen's lives at risk.

"No," Barði shook his head emphatically. "The water absorbed the shock. Any whiplash effect in the trawl wires went sideways, not toward the trawler. There was no danger to the trawler. "But, if the cable to the cutting gear snapped," he paused for effect, his face solemn, "the whiplash would come straight back onto the deck of the gunboat. That made it riskier for us."

Barði showed us the self-steering gear in his wheelhouse, and explained how he used the echo-sounder to locate shoals of fish.

"Cod usually stay in the same area, so once I've found a good place to fish, I use the radar to pinpoint my position. That way I can easily find the same spot again.

Iceland by Kayak

"Sometimes I follow other ships if I know they are catching fish. Mostly, I stick to the patches I know best, around Langanes, Tjörnes and Grímsey. Just occasionally I go around Horn, but that makes for a long trip."

We left Barði when his turn came to unload, bought food at the shop, and returned to the tent to cook. There, Barði's little daughter, Helga, stood and watched us in silence, having followed us into the tent. As soon as we had eaten, Helga and her friend Gunnur invited us to Litla Heimilið, *little home,* their small wooden play shelter in the garden. There, with formal hospitality, they served us sand pudding, with teacups of muddy brown water.

Windy weather afforded an opportunity to write letters, but shopping for picture postcards left us disappointed. Expecting to find cards with Icelandic landscapes, our options only portrayed dogs, just one card showing a map of France with a briar pipe resting on it.

We returned to *Trausty*, Barði's boat, in time to hear the updated weather forecast. "At the end of Melrakkaslétta, expect winds from the northwest at force four to five, decreasing." Melrakkaslétta was the broad peninsula we looked forward to rounding next.

We were glad to hear the winds should ease. In the meantime, Barði invited us to coffee with his wife Anna, who was a joy to chat with. She credited her fluent English to time spent with her aunt in Nottinghamshire, England.

In the evening, when Anna and Barði came to see our kayaks, other boat skippers gathered around too. With little to show, we asked if anyone would like to try.

The first man, pushing powerfully from shore, capsized at once. Undeterred, he fared better on his second attempt. Emboldened, Barði tried. He wobbled from side to side, sometimes fending off the bottom to balance, bracing wildly from side to side until, trying to reverse, he fell in.

Fox Plains

Others clamored to take a turn, despite each capsizing at first, for our kayaks were especially tippy with no cargo. Most succeeded on their second or third try. The rest of us cheered them on, shouted encouragement and laughed heartily.

The evening wind proved a cruel goad to the men in wet jeans and sweaters. Their fun was brief. Saturated, one after the other hurried home. Alone at last, we drained our kayaks, lifted them ashore, and called it a night.

Next morning, brave in a t-shirt, I savored the sensation of sunshine on my skin. Recalling the start of our trip, when it never grew dark, I tried to remember how warm it had been. Surely, as the sun retreated, the nights must grow cooler, summer edging away.

With the thermometer reading five degrees Centigrade, (41°F), Anna came out to tell us the weather forecast sounded good, with a southwest breeze expected later. In the house, she poured us coffee and made us sandwiches, saying with a smile that Barði had enjoyed his escapades last night. "I found it amusing too," she added, smiling wider in recollection. Then, she announced happily, "He's gone fishing today."

Paddling north we had the visual stimulation of cliffs to follow. Caves, arches, and stacks embellished these cliffs. It was the presence of both columnar basalt, and softer rock resembling hard-packed ash, which created the natural architecture. The softer rock had worn into smooth, pillow-like surfaces leaving the sharp-edged, angular basalt defying the erosive force of the sea. Dark pillars of that harder rock splayed out in all directions.

By the time we reached the northwest tip of Melrakkaslétta, the breeze had increased considerably. We tucked ourselves under a cliff of grey and brilliant red rock for a short break. Horizontal bands of hard rock jutted out as narrow shelves. Beneath the grass overhanging the top of the cliff, the earth glowed orange. Anticipating rougher seas around the headland, we donned the

Iceland by Kayak

anoraks we had foregone to start, and were glad, for beneath the lighthouse, swells reared up and broke, rolling across the shallows surrounding two massive pillars of rock. Beyond that, the sea looked energetic, frisky. Was it the wind opposing the tide roughing up the water?

The wind blew from the east, in our face, not from the southwest as predicted. Headwinds always slowed and soaked us. The kayaks pitched up through waves, lurched over and dived. Each time my bow plunged, water swept the deck to slam me in the chest and face. Cinching the hood of my jacket, however tightly, never prevented water from gushing down my neck. Water forced up my sleeves, draining from the armpits to the cockpit.

As our cockpits filled, water slopping deeper around our legs, we rafted together to bail. Steadying one another, each in turn grasped the red and black handle behind the other to vigorously pump out the water. I had ill-advisedly positioned my pump outlet on top of the deck just behind me, so bailing sent jets of water skyward, showering me. Geoff spared me, mostly, by resting his paddle across the outlet to divert the flow.

Rafted together, our kayaks inevitably swung side-on to any wind while we bailed. Waves hitting from the side washed over our decks, while the wind pushed us sideways, robbing us of hard-earned ground. Uncomfortable, awkward, and frustrating as it sometimes proved, the reward of a more stable kayak with a warmer, drier cockpit made up for it.

The sea was not getting easier. At Haganes, we tried to avoid an area of breakers by detouring out to sea. Too complacent, we misjudged how far offshore to go. Partway around the headland, we mistakenly strayed among rollers. Geoff and I spread apart, lest a wave threw one onto the other. Suddenly, a steep wall of water pitched forward and roared down, the breaking water narrowly missing me as I lurched steeply up and over, breathless, onto a hissing surface of rising bubbles. Rainbows hung in the spray. The roller pounded onward, thundering toward the rocky shore.

Providentially it had missed us both. The alternative, someone carried onto the rocks, or to avoid that, swimming, would have been a disaster. Intimidated, I tried to dodge the steepest waves which peaked abruptly and broke unpredictably. Working ever farther offshore, at any moment certain to be in the wrong place, it was like Russian roulette. Skill was not enough; I needed luck. I was relieved to reach open water without disaster, and to see Geoff emerge safely.

If not for that drama, we might have paddled farther. Instead, we sneaked into a small bay just beyond Haganes. Finding it quite protected from the waves, we landed near a house at the end of the beach to fill our water bottles. The house was unoccupied, so we scooped water from a stream.

Eyeing a steep-sided grassy hollow slightly smaller than the tent, we pitched the tent over it. The bank inside offered a backrest to lean against and afforded more headroom. While Geoff fired up the stove beside the kayaks outside, I leaned against the bank inside and stretched my legs. "This is deluxe," I called, comforted to hear the hiss of the stove steady to a roar.

"Nigel, could you pass out one of those dried fish, the ones Guðmundur gave us. The bag should be in there somewhere, at the back of the tent."

I located the plastic carrier bag from Flatey Island. Recalling how the fish were tied, tail to tail, to drape over the boat rail to dry, I opened my knife to cut the line. When I reached into the bag to grab a fish, I recoiled, jerking my hand back out in shock. The fish had tickled my palm. Gingerly, I spread open the bag. The fish writhed with fat white maggots. I shivered, my hair prickling. Disgusted and queasy, I rolled the bag closed again. In a moment of mischief, without a word, I passed the bag out to Geoff.

Iceland by Kayak

Still jittery from the shock of opening a bag full of maggots, I settled back, listening. Moments later, Geoff swore loudly as he flung the fish far out to sea. Neither fish, nor maggots, were on our menu today.

We launched early next morning in short-lived sunshine. By nine, mist closed in, and everything vanished, forcing us to focus closely on the map and compass. Consequently, the sudden appearance of land just ahead always startled us.

"Everyone a winner," Geoff retorted, making me look up again from my compass. There was a shipwreck right there. Beyond it, gauzy in the fog, lurked a point of land. We crept past the wreck and headland, into fog again, crossing blindly in the direction of the next point. There, more wreckage appeared, this time a huge rust-red boiler perched on the rocks. "Everywhere seems to have its own wreck! Unless all this came from one ship, it must be a tricky coast."

The mist dissipating, we rounded the northernmost point of mainland Iceland, Rifstangi, the closest to the Arctic Circle. An hour later we passed a tall, square lighthouse at Hraunhafnartangi, the northeast corner of Melrakkaslétta. Having worked our way around the peninsula, north, then east, finally southbound again we paused at Raufarhöfn to refill our bottles from a dock hose.

Raufarhöfn, a village in a natural harbor, was once one of the largest herring stations in Iceland. It had since become quiet. There, a fisherman on the dock gave us a fish. "The red ones," he pointed into his tub, "eat red worms, and that colors their skin."

The blunt peninsula, Melrakkaslétta, *fox plains*, consists of a heart-shaped plain nestled between the eastern and western arms of mountain ridges that slope down northward into a Y-shape. This wild area is known for nesting birds, hence foxes. The arctic fox, a survivor since the last ice age and smaller than Britain's red fox, is Iceland's only native mammal. We soon reached the eastern arm of higher land, and its northeastern point, which juts

Fox Plains

out as Melrakkanes, fox's nose. There we quit, establishing camp on a grassy shelf below a crag which rose steeply to the headland. A ramp of debris formed a smooth transition, a green slope, from the near-vertical face to the flattish coastal shelf.

This place made me apprehensive. Large boulders lined the shelf, and it was obvious how they got there. Each had gouged a deep furrow down the gradient, stopping just short of the ocean. The deep-cut scars were as wide as a room. Some of the furrows were grass covered, but exposed bare brown earth showed others to be recent. Such freshness concerned me. The risk of a rock hitting us was small, yet the consequence would be disastrous.

In the south of England, training with my friend Jan for our first crossing of the English Channel, circumstances forced us to bivouac overnight on the clay and sandstone cliffs of Fairlight. The weather forecast promised fine weather, but we got torrential rain. Throughout the night, I heard occasional rolls of thunder. Both of us drenched by morning, our priority was to boil water for coffee. Startled by a roll of thunder, we turned our heads to watch a slab of sandstone trundle down the cliff. The downpour had softened the clay, weakening the support for the layers of sandstone. There had been no overnight thunder, only rockfalls. Foregoing coffee, we left in a hurry. Would there be rockfalls here while we slept, I wondered?

Our next morning's sights set on Langanes; the windy, rainy conditions deterred us. Roaming at a loose end, we began to find puffballs and began hunting for more. Sliced and gently fried until golden brown, these edible fungi had a flavor and texture resembling omelet.

Our tent was tired, worn thin by the winds. To escape the squalls, we huddled in its drier end. Whenever it rained, water dripped heavily from the worst places in the stretched fabric and let in a spray of fine mist everywhere else. It was the result of two months of harsh treatment.

Iceland by Kayak

Of course, the tent was not the only thing to show signs of wear. Our anoraks leaked profusely at the armpits, where we suffered each intrusion acutely. Our shoes were pitiful, the canvas having rotted. Sand had long ago worked its way inside my camera. And of course, there was the stove. Apart from all that fair wear and tear, our gear was holding up well.

During a brief break in the rain and hoping for a better view we hurried to the lighthouse, huddling in its doorway while the next squall blew by. The sea was lumpy and the wind blustery. We were both restless, so between showers we took to picking blueberries and searching for puffballs. From the beach we scavenged a sheet of polythene, to cover the back of the tent against the rain.

When the sun sank low, we climbed the unstable cliff to view the sunset. On the top, the ground was gravelly, sparsely vegetated with low arctic-alpine plants. Boulders perched where erosion left them teetering, almost ready to topple. Those near the edge looked precarious, yet the monoliths along the shore appeared to have come from the cliff face, not from here on top.

Relentless squalls continued into the night. Under each assault, the tent rattled and punched, and a loose corner of the polythene sheet slapped against the tent by my head. Alert, I listened for the thunder of rockfall. My hair felt damp, as did the neck of my sleeping bag. Unable to sleep, to distract myself I thought about my new job at Burwash Place.

I pictured the old stately building, its long tile-hung frontage and latticed bay windows, glowing in the morning sun. I imagined the Virginia creeper, encasing one end in well-trimmed foliage, blazing autumn red after my return. Large sandstone slabs paved a wide path the full length of the building. That weathered sandstone, from a local quarry, split along its bedding planes to reveal ancient ripple marks, pleasant beneath bare feet.

Fox Plains

Shallow steps dropped to the broad well-kept lawns, and beyond that ran a hedge that hid the valley. The far skyline, the sandstone summit ridge beyond the clay valley, defined an east-west trading route already in use by Neolithic times. That was long before the Romans invaded and took advantage of it, which in turn was long before anyone settled in Iceland. Straddling the road that still follows that ridge lies the ancient village of Burwash, where parts of the church date back to the eleventh century.

Recalling Burwash Place, I pictured the lawn again. To one end stood a magnificent Lebanese cedar tree. It spread dense fan-like layers of foliage, like giant blue-green hands held palm-down. A lightning strike, that once tore away part of this great tree, left its huge curving lower boughs. Preparing for this trip, I had considered climbing into those lower boughs to sleep. An earlier instructor, Jack Grasse, had done so in jeans and a T-shirt, to harden himself against the cold, in preparation for an expedition to northeast Greenland.

Late one night, I crept barefoot across the dewy grass to that tree. Ready in the dark with my hand against the rough bark, I talked myself out of copying him. It was not for fear of falling while asleep, which did cross my mind. No, it was the spectacle I would become in the morning when the students looked out, and the bad precedent I might set. This tree was out of bounds to the students. Besides, since Iceland had no trees to climb, my motive would seem absurd. I imagined the scenario: "What are you doing up that tree?"

"Oh, I'm getting ready for a kayaking trip to Iceland."

No, I decided, the tree could wait. Besides, the Norwegian explorer Roald Amundsen prepared for the arctic by sleeping naked by an open window in winter, a harsher preparation.

Tonight, lying here in my sleeping bag, despite the rattle and slap of the tent and the dampness, I could admit that summer in Iceland was not that uncomfortable.

SKORUVÍK, LANGANES. CAMPING WITH DRIFTWOOD.

GEOFF BELOW BASALT SEA STACK.

Iceland by Kayak

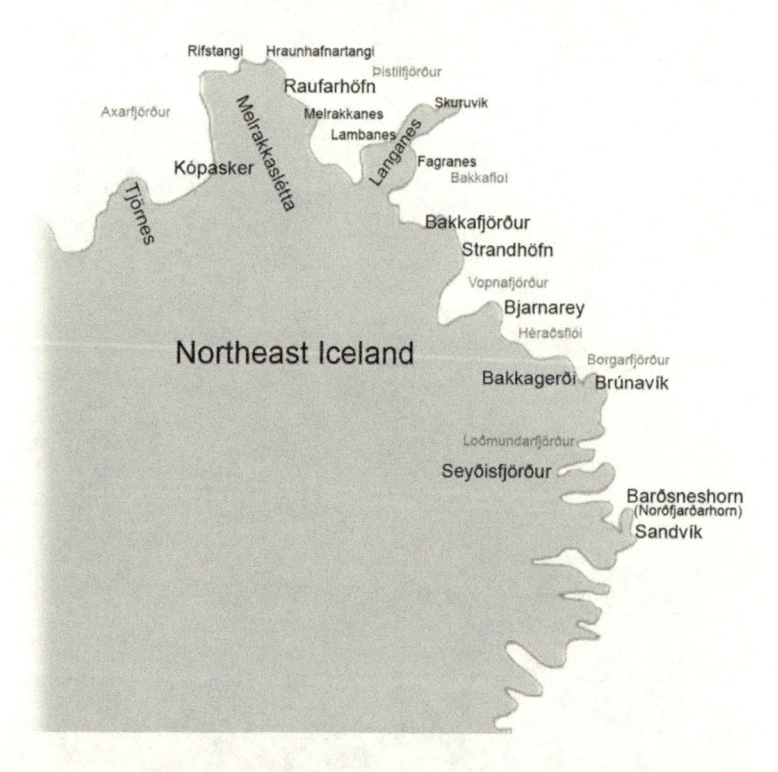

MAP 10. NORTHEAST ICELAND.

28

Langanes

Of the four areas I was most apprehensive about when first planning to circle Iceland, we had passed three. I had predicted the sandy south coast would be the crux. Coastguards, and other people who should know, warned, "You will not be able to pass," or, "Skip that part: it's too dangerous and remote. There's brutal surf with dumping waves, sandstorms, and quicksand." I was glad we tackled that first.

Next came Reykjanes promising currents and ocean exposure. The headlands of the third, in the northwest leading to Horn, threatened tide races, with potentially heavy seas.

Now we approached the last, Langanes. Pointing northeast like a witch's knobbly finger, this was where the meeting of currents rushing offshore might cause a rough sea. Beyond Langanes lay the remaining one hundred miles or so before Seyðisfjörður.

So far, when close to the coast, neither current nor tide had amounted to much. Not in a hurry, we could afford to watch the weather and choose the best time to negotiate any awkward area. So, I saw no special reason to expect problems here. Moreover, I looked forward to a psychological boost once we turned the

Iceland by Kayak

corner, from the north coast to the east, into the home run toward Seyðisfjörður. In passing Langanes, we would also cross from the ancient administrative North Farthing, our fourth, to the East Farthing, the district in which we began and would finish our trip.

The name, *farthing,* reminded me of the British coinage of my childhood. Four farthings, *fourth-ings*, made one penny. As small denominations fell out of use, the farthing ceased to be legal tender in 1961, followed in 1969 by the halfpenny, *ha'penny,* worth two farthings. I still have a Queen Elizabeth II farthing, minted with the image of a wren on the other side.

I also remember, when I was little, reading handwritten price signs, square cards on sticks, with prices like "3¾d", pronounced *thruppence three farthings*. That in a store where a bare-armed woman, in a stiff white apron, batted a mound of butter into a tidy block using two wooden paddles. She sliced the butter into bars with a wire, to weigh, before wrapping in crackly greaseproof paper. There were large trays of eggs, and giant blocks of strong-smelling cheese.

The "d" for pence puzzled me. A spelling mistake, or written upside down? Years later, in Latin class, I learned how the penny came from a Roman silver coin, the denarius, hence the symbol "d." Nowadays, children need not puzzle. UK decimalized its currency in 1971, denoting new pence with the symbol "p." [5]

Farthings aside I felt proud, viewing the map, to see how far we had come. With practice, I had committed our watery route to memory as a string of landmarks. The once bewildering placenames now evoked islands, bays, sands, and headlands. My tongue could articulate the topogeny, like a poem: *Seyðisfjörður, Sandvík, Höfn, Ingólfshöfði, Efsta-Grund, Vestmannaeyjar, Þorlákshöfn, Reykjanes, Keflavík and Reykjavík, Akranes, Faxaflói, Snæfellsjökull, Breidafjörður, Suðureyri* and so on.

As each rolled from my tongue, I visualized its location and what we had experienced there. I could detail our route on the

Langanes

chart or scribble a pencil-on-paper outline of Iceland, to pinpoint details when quizzed. The oral repetition helped me to memorize and streamline our story.

To my surprise, Icelanders always seemed hungry for details, their curiosity genuine because everyone knew places we had seen, and people we had met. In a land of so few, the net of relatives and acquaintances casts wide. It was important for us to remember names.

So, onward to Langanes, across twenty miles of Þistilfjörður, *thistle fjord*, to the peninsula. We began once the weather began to clear and the wind dropped sufficiently, despite the sea remaining active. We had passed one fishing boat hauling its net, and had paddled for more than two hours, before I became vexed by Geoff always keeping slightly ahead of me. Whenever I paddled faster to try to catch him, he sped up, keeping ahead. When I slowed, he slowed too. Was he unaware of what he was doing? I appreciated his not forging ahead as he surely could, but it bothered me to be always just behind him.

During a long crossing, with little to gauge progress by, it is natural to aim for a distant peak or cliff, to watch the transit points shift incrementally. I am content to recognize progress hour by hour, despite seeing negligible change over ten or fifteen minutes, but that was impossible with the stern of Geoff's kayak just ahead. Instead of watching and focusing on the distant transits, I irrationally measured my progress toward the obvious milestone of his kayak. Powerless to gain, no matter how hard I tried, I suffered under the impression I was not making progress, a psychologically negative illusion.

Geoff knew about this phenomenon. We talked about it early in the trip. He had trailed behind other kayaks before and knew how it felt. So, was this deliberate or inadvertent? Had I upset him, or was he unaware? Frustrated, I kept it to myself.

Iceland by Kayak

A cloud crossed over us, with a chill squall of wind and more energetic waves which eased again once the cloud passed. Later the wind picked up, blowing strongly into our face. I knew how Geoff disliked paddling into the wind, while perversely I quite enjoyed it. As the sea got up, and the wind increased, I now pushed especially hard to get ahead of Geoff and endeavored to stay a little forward of him.

Empowered by imagining my strength in these conditions, I had the psychological edge. Typically preferring to collaborate rather than compete, I felt different today, determined. The sea grew rougher still, as the wind continued to build. The kayaks launched up and sliced down through the oncoming crests, dousing us over and again, water draining into our kayaks.

Stopping to bail, clutching our kayaks together for balance, we considered changing course. Would a closer target be prudent, one that also promised respite from the direct headwind? Looking diagonally across the wind, Lambanes, *lamb point*, would be an easier target, a better angle through the waves. At least that is what we thought.

The still-strengthening wind continued to veer, turning to blow sideways across the waves. Spawning new wave patterns across the old, it lumped up the sea, generating watery chaos. It was not long before it drove directly at our face again.

Forging on anyway, utterly drenched, we at last reached Lambanes. With the momentum from a final sprint toward the flat yellow sands, our kayaks slid from the water before coming to rest. There we sat, in silence, for a long moment, soaking up the sun's warmth. Beached, in no hurry to get out and stand up, we laughed at the effect of the wind blasting first cold, then as hot as from a hair dryer.

"This feels so weird. Remember that day in the southeast, when the winds did just this, right at the end of that long day?"

"Ugh, of course! How could I forget?"

Langanes

All my earlier frustration long since dissipated, I felt a twinge of guilty pleasure when I heard Geoff declare this the worst, most uncomfortable crossing. He also pointed out that, currently low on food, his boat was heavier than mine. He thought he was losing speed with that.

I said nothing, still considering him the stronger paddler. Excuses were unnecessary. It was satisfaction enough for me to imagine I had scored a psychological victory however small, valid, or not. It would no longer bother me if he kept ahead.

After our break, we left the sand and hugged the shore until we reached the lee of cliffs. There, we could paddle more easily and pleasantly. Occasionally a violent downdraught caught us unawares, racing across the water from one direction or another, ripping up a frenzy of capillary waves, sometimes a burst of spray.

As the cliffs grew steeper, and taller, more nesting seabirds appeared, and the skies became alive with action. Kittiwake nests plastered the steepest, most precipitous rock faces, where it seemed implausible that they could cement a nest to the rock at all. Once the chicks in the nests were fully fledged, what would they do next? Spread their wings and take to the air faultlessly, as if they had been learning to fly from the moment they broke from the egg? Or plummet? Judging by the sodden carcasses floating beneath the cliffs, that first excursion from the nest is a steep learning curve.

We passed a colony of gannets, densely crowded and dazzling white, covering the top of a large stack. Oversized, the gannets stumbled, and stabbed, and croaked in a dizzying turmoil, like creatures from an earlier epoch. Beside them, even the biggest seagulls seemed diminutive.

I gazed in delight as a gannet drifted past on the air, majestic with its vast, black-tipped wings held aloft. At the last moment, its aerial poise ended with a stall and fall from grace. The bird, with wings akilter, scrabbled for a foothold amid the long thrusting bills of those sitting on the closely grouped nests. Fumbling, and stumbling, trying to flap toward its nest, it braved a gauntlet of spiteful beaks. All this amid a cacophony of grunting, grating calls, and the hollow slap and burst of wave against cliff. The air clawed thick with the pungent smell of guano.

This section of cliffs formed shapes that jutted seaward like the upswept bows of ship, after ship, moored side by side pointing out to sea. Bird lime, green slime, and verdigris stained the faces. The wind gusting over the edges, sprayed dirt, eddying it around and down. Both wind and debris frilled the water surface, while clumps of lichen, and feathers, soared and eddied. The crash, thump, and wallop of wave against rock, the hiss and splash of spray, the shriek and grunt of birds, all echoed, amplified within the confines of the cliff faces.

With me gleefully just ahead of Geoff, we reached Skoruvík, *score bay,* where we landed on a beach piled deep with driftwood. Stumps, logs, and spars jumbled high all along the top of the beach and up against the low cliff. This was the last suitable place to stop before reaching the tip of Langanes.

When Geoff admitted how tired he felt, I again said nothing. Psychological, I surmised. Had he kept just ahead of me, as usual, I would have been the one feeling tired instead.

"It seems as if Langanes doesn't want to let us pass," he complained. "It's already held us back for days, and the closer we get the more belligerent it gets. The wind keeps swinging around in our face whichever way we turn."

We trudged the first kayak up the gravel beach, to where a flat berm of boulders disappeared beneath the tangle of driftwood, scrunching back for the second. The misty grey sky leached far lower than where the horizon should have been.

We set to, clearing a space for the tent just above the tide line, and got into dry clothes. With no stream visible for fresh water, we balanced carefully from log to driftwood log, to reach higher ground with a wider view. Beyond the beach and the crumbly cliff, a dirt road ran parallel to the shore. Beside the road, a little way along, stood a single house.

Approaching confidently, we knocked on the door. A dog barked, and I heard a voice. Someone was home. Presently, the door opened a crack and an old man peered out at us. We held out our bottles, asking for water. He understood, although he seemed not to speak English.

He kept the door just partly opened, standing mostly hidden from our view. His dog cowered behind him, not blocked from escape. It was as if it were afraid to run out to greet us, sensing danger, held back by fear rather than obedience.

The old man shouted behind him into the house summoning two women who joined him at the door. I assumed they were his wife and daughter. Unsmiling, they reached their hands past him to snatch our containers while he continued to guard the door.

When they returned with the filled bottles, he thrust them out, one at a time, through the crack in the door and closed the door.

I stared at Geoff. Like me, he stood damp and salt-encrusted, wild-haired, and sea-pinched. He stared back. The arrival of such strangers here must have been scary. How should the man have reacted at the sight of two vagrants at the door with no visible means of transport? Yet as we trudged back toward the beach, Geoff broke out laughing and exclaimed, "That was amusing." He mimicked the scenario in a short monologue, "Quick Rita! Load the shotgun! I'll hold them at the door for as long as I can!

"And then there were all those footsteps we heard running around."

Now, in a higher voice, "I can't find the shotgun cartridges! Oh dear, where did you put the cartridges?"

Then, in his own voice, deep with menace, "Never mind the coffee old man. Just fill up these plastic containers with gold and nobody will get hurt." Geoff swung his water container with glee.

Finally, imitating a newscaster's voice, "After a daring daylight gold robbery at Langanes, the thieves appeared to vanish like ghosts into the beach!"

I laughed. "Well, I don't suppose they get many visitors way out here. There's no road, just that track that goes to the lighthouse. The last farm was at least ten miles back. They must wonder where we came from. Still, I am thankful for the water."

Back on the beach, Geoff once again scavenged part of a rusted oil drum and built a fire inside it. From the flotsam, we disentangled a stout hawser from a half-buried net. With boulders piled to support two logs upright, we strung the rope, hung our gear, and sat back to enjoy the fire, replenishing the flames with driftwood from within arm's reach.

So far on our journey we had seldom seen a tree, and only diminutive specimens. Ocean currents affect the coastal timber line. In Norway, warmed by tendrils of the Gulf Stream, the timber line lies at latitude seventy degrees north. In Labrador, cooled by the Labrador Current, it runs 850 miles or so farther south, at around fifty-six degrees north.

Langanes, north of latitude sixty-six degrees, cooled by the East Greenland current, must lie close to that natural boundary. Did that make it easier for sheep to prevent reforestation? For more than a thousand years, those multicolored sheep had grazed down any tree seedlings.

I began taking stock of driftwood. Piled deep, it was not the same mix we had seen on the beaches farther west. True, here was the same felled timber, Siberian logs, the ends rounded like worn erasers on giant pencils. Long since stripped of any bark, the surface grain of these logs had splintered into a fuzz of soft, salt-soaked, hairy wood fibers. The polar ice had carried these drift logs.

Scattered between, and pinned beneath the logs, were boards, squared timbers, and spars that might once have been parts of boats, or buildings, debris of the kind that might have come from the decaying docks at Siglufjörður. And here and there, as always, industrial fishing floats, and tangled net. The currents, waves, tide, and wind had been selectively generous to this beach, while scouring others clean. People might have named this beach Keflavík, *driftwood bay*.

So why, I wondered, was it called Skoruvík rather than Keflavík? Skoru means *score*, but in which context? In old English, dating back a thousand years, people counted sheep in groups of twenty, scratching or notching a line on a stick for each twenty. Nowadays, a score means twenty. You mark down numbers to keep tally or *keep score*. Was there perhaps a connection here with tallying sheep?

When you deliberately scratch a stick, you score it. You make a surface incision, indenting a line into paper or wood. In Icelandic, *skoru* means score, and *skor* means a rift, precipice, notch, or incision. Here at Skoruvík there is a notch of sorts.

A low straight valley-line, called Vatnadalur, scores the land. This runs between Skoruvík and Skálar, *the basin*, an abandoned settlement little more than two miles away on the southern shore of Langanes. The map shows a bridle path following that score line.

Whatever the origins, tallying sheep, or an incised landform, when I retired to the tent with the sting of salt and woodsmoke in my eyes, I had no need to count sheep.

***

Powerful downdrafts still buffeted around the low cliffs when we next launched. Such blasts could foretell strong headwinds on the other side of Langanes. Alert with anticipation, hugging the shore, we slipped easily around the point into warm sunshine and only a light breeze. Clean swells ran in procession under a blue sky that appeared set for the rest of the day. Grinning happily, we shortcut the bay toward Fagranes, and aimed across Bakkaflói, *back bay,* toward Bakkafjörður.

The conditions were too good to last, and when the wind sprang up, it spawned an erratic, jumpy, jostling sea with peaks of water sprouting all around. It was the uncomfortable result of the local wind-waves at odds with the prevailing wave pattern, colliding and leaping. The sun dazzled as the erupting peaks flung out plumes of sparkling jewels to drench us.

Our kayaks speared into and reappeared from under breakers. They leaped upward from each crest, hanging for a moment before plunging, axe-like, to cleave the water with jarring impact. The pitching action slowed us, yet little by little we crept close enough to Bakkafjörður to line up the lighthouse against the cliffs beyond. To our right, we monitored our progress against a headland, Stapi. Gradually, amid blasts of balmy air, we drew close enough to select a part of the village to aim for, finally cruising past moored boats to the beach.

The moment we touched land, a woman with great posture and a pretty face came strolling down the beach to greet us. She introduced herself as Kristbjörg.

"You speak English well!" I complimented her.

"Thank you." She explained, "I spent one year studying English at Bournemouth ten years ago, before I married."

Another girl hurried to join us. "Hi, I'm Pýrleif," she greeted, "I work at the shop. It's closed for the weekend, but if you like, I can open it for you, if you need to buy food." She looked us up and down. "But please, you should change first. You are wet through. Aren't you cold?"

Kristbjörg's husband Hörður had joined us by the time we reached the shop. He said they were visiting from Reyðarfjörður, farther south on the east coast, beyond Seyðisfjörður. They were staying with his sister Ingibjörg, Pýrleif's mother. Hörður invited us home for coffee where we met the rest of the family.

Langanes

"Take a shower if you wish," Hörður gestured. We leaped at the chance to rinse off the salt.

Pýrleif's brother Birgir played guitar. He fetched his instrument, and we began to exchange songs. Meanwhile Hörður, Kristbjörg, and their son, left on a boat to go fishing for a couple of hours. They returned, content, with a catch of about fifty cod.

"You must take one," Hörður insisted.

By the time we left the house to find our kayaks and set up the tent, already one o'clock in the morning, it was almost dark.

Next day, when the sun grew hot enough for us to bare our arms in t-shirts, Birgir brought his guitar to the beach. He asked me to teach him the George Harrison's song, *Here comes the sun*. He learned quickly, practicing while I loaded my kayak. Once he had mastered each new part, I showed him the next.

Birgir's friend, Hilmar Þór, a singer, observed, "I expect you only see the coast. Everything is quite different when you go inland. If you like, I'll take you for a drive."

The first difference I noticed, heading inland beyond the end of Bakkafjörður, was that it was hotter. We parked the car and climbed a hillside to a massive clump of bilberries, where we sat down to pick. The juicy berries were fatter, and sweeter, than any we had collected on the coast. The warmth of the sun on the hill brought out the aroma of sweet cinnamon, and the comforting mumbling murmur of foraging bees.

Driving again, we passed a house where a girl lay out sunbathing in a bikini. Þór pulled up abruptly and spun the car around, to drive past at a crawl, to look more closely. Then, he turned once more and crept past to ogle her again. "She's a visitor!" he protested indignantly when we teased him.

"How do you know? And what difference does that make?"

"I know all the girls in Bakkafjörður," he explained. "I've grown up with them. They are all like sisters to me. Like family. This one, well, she's different. She's a visitor. A novelty." Then, he added with a grin, "And, she's wearing a bikini." He was thrilled.

It struck me how few girls of his age must live in Bakkafjörður, with its no more than sixty inhabitants. It was to his disadvantage, growing up in a small settlement.

Geoff and I left Bakkafjörður that evening, once the wind began to ease, and paddled around the headland into fog. We had chosen to paddle to Bjarnarey, *bear island*, but darkness spread as the dank fog thickened. Cutting the day short, we stopped after about fifteen miles at Strandhöfn, *beach port*. There, we surfed between rocks onto a steep beach.

Shivering on the beach, eating pudding, discussing where to put the tent, Geoff observed, "This has all the qualities of an English winter paddle."

"Miserable, cold, damp, and dark?" I asked, with a grin, knowing full well what he meant. "Winter? Ha! Well, just remember that girl sunbathing in a bikini today."

We pitched the tent on flat ground at the top of a low cliff. With a farmhouse in sight, we set off to ask for water. The farmer invited us in for coffee.

"We're hay-making," he explained at the table, by way of conversation. "We have a hay-liner," he added.

"It's a baler," said Geoff, who must have recognized my blank look. "They don't need to do all the work by hand like at Gjögur."

The farmer continued, "We keep four hundred sheep. In winter we house them in sheds, so it is easier to manage the hay in bales. We also have two cows and a heifer, and chickens."

It was warm, so with hot coffee and snacks, my salt-sore eyelids grew heavy. There was something wonderfully comfortable about sitting in a chair.

ROCKS SECURE OUR TENT.

CLOUD-CAPPED MOUNTAIN, EAST COAST.

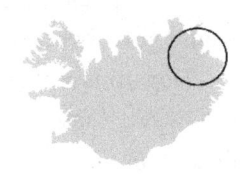

29
East Coast

When the whole farming family came to the beach next morning, curious, they pitied us in our sagging tent. I saw the reason. It drooped, damp and limp. Stove soot grimed the weather-bleached nylon. Where the fabric had rubbed against the frame and worn away the waterproofing, the dark stain of aluminum impregnated the weave.

"Come back for coffee," they invited, and with the coffee, came a generous breakfast of boiled eggs, sandwiches with cheese, tomato, and cucumber, and cake.

One of the men, a visitor staying there, told us proudly, "I have a farm in the south. I grew those cucumbers and tomatoes there," he pointed at the food on the table. "We also grow bananas."

"Bananas?" I was incredulous. "They're tropical! How is it warm enough?"

"Greenhouses. We heat the greenhouses with water from the hot springs. We have twenty-four hours of daylight in summer, so with the heat, everything grows very well." He bagged tomatoes for us to take with us.

Iceland by Kayak

"Tomatoes, yes, but bananas?" It seemed so unlikely they would grow here. Was he pulling my leg?

Departing, we pushed against our nemesis, the daily headwind, across Vopnafjörður, *weapon fjord*, passing a fisherman who was handlining. We gratefully accepted the first two cod he offered, declining more.

Landing for a break on a small, sheltered beach, we collected pebbles of all colors. It was amazing to me to consider how, from their igneous origins, the rocks had worn smooth and collected here, each pebble so differently patterned and colored. Vibrant shades of yellow, bright red, green, all born of fire.

Ahead as we passed Bjarnarey Island lay Héraðsflói, a broad bay backed by a sand beach with a low valley beyond. Weary of the wind, we hugged the cliffs to enter the bay, past tremendous sulfurous canary-yellow walls, and jagged pinnacles. Folded veins of hard rock, volcanic intrusions, sprouted from the cliffs.

Within the bay, the surf forceful, we kept well seaward of the break line all the way along the fifteen-mile beach, not reaching the next headland until it was almost dark. By then, we thought we could make it to Brúnavík, *brown bay*, a small bay, the next east past Borgarfjörður. The Admiralty Pilot promised a refuge hut there.

Windless, all was quiet when we crossed Borgarfjörður, creeping at a distance past the town of Bakkagerði in semidarkness. I could see the mountains, and the twinkling house lights at the head of the fjord. The summer night skies had not yet fully surrendered to winter darkness. Although it was too dark to read the details on my chart, I could still discern the colors of Geoff's kayak.

My arms grew weak as I realized Brúnavík was farther than I had expected. Those few extra miles grew longer, until I fell under the illusion we were paddling on the spot, not moving forward at all. My mental torment compounded my physical weariness. My arms grew leaden, my willpower diminished.

Yet, we did eventually reach Brúnavík, where a stream cut through the beach. There, in the gloom stood the hut. Lighting a lamp inside revealed bunk beds, and a table. It offered the warm welcome of a home.

Fetching only what I needed, I quickly settled to sleep, my legs dangling over the end of a short bunk.

Geoff's relaxed voice came from the darkness, "You know, I quite fancy taking a couple of days to finish the journey to Seyðisfjörður."

Having paddled more than forty miles today, there were only twenty-five more to go.

"I'd prefer to try to make it there in one day, if the conditions are okay," I countered. "You never know how long we could be hemmed in by the wind, if we wait."

Geoff muttered something like, "What's the big rush." I drifted off to sleep. In my mind, I was already tasting the joy of completing what had so often seemed an endless undertaking. I was excited to finish.

On waking, we threw open the doors to a beautiful morning, the sun dazzling. Fences ran close to the hut. They made it feel like a farm even if, save for this building, there was nothing in sight. Brúnavík was a bay surrounded by mountains, with grass in the valley. And these straggly fences, were they for rounding up sheep? It must be difficult to get here overland, otherwise why build a rescue hut?

We lazed in the sun, flicking away flies, writing, and boiling a suet pudding, tending our paddling garments as they dried. Geoff came across a puffball and cooked the tasty morsel. With no sense of urgency, I had become invigorated by being so close to the end.

I recalled how it all began, at Dave's house, with the atlas inset map of Iceland. Months of planning had followed, before our drive north began with our newly designed kayaks. Every part of the process had been an adventure. We had hoped to paddle all the way around Iceland, not knowing if we could. Now, it seemed inevitable we should succeed.

It was a leisurely two-thirty in the afternoon before we launched onto a calm sea, only to find, to Geoff's annoyance, a brisk headwind waiting just beyond the shelter of the headland. We had countered enough of these contrary winds to last him for a while. Besides, he had voiced his preference to wait another day before leaving.

Spellbound, we passed the most magnificent basalt formations. Here were menacing dark cliffs of trunk-like basalt columns, contorted formations, clusters of curving pillars splayed like fingers. In places, horizontally aligned columns had eroded into walls of hexagonal blocks, so regular in shape they looked fabricated. I might have believed they were, if told, except for the massive scale, and the location. Bouquets of dark grey curving columns blossomed from the cliff faces, sometimes weathered unevenly into embossed whorls. The effect was reminiscent of spiral stairways boring sideways into the cliff face, the black patterns like monochrome wood-block print. Waves clunked and walloped against deep-cut corners and hollow alcoves, the dazzling foam draining from wet-darkened rock.

Ahead, an over-spilling roll of cloud obscured the top of a mountain ridge jutting out to sea. It reminded me of the one we saw earlier in the summer, at Eystrahorn in the southeast, pouring over the ridge and dissolving into clear air. Here, as it fell, waterfall-like, I feared similar fierce winds. Fully expecting alternating blasts of cold and hot wind, I was surprised instead by an embrace of chill moisture before mist engulfed us. The sea calmed into a deceptively sleepy swell, which steepened without warning. As the mist clenched into fog, it demanded our wits to safely weave a path between rocks, stacks, and ledges, avoiding boomers that thundered out of nowhere.

When at last we reached Loðmundarfjörður, *furry-mouth fjord*, the visibility cleared. To my delight, I recognized the same mountain ridges we had seen from the deck of the ferry, Smyril, when we first arrived. There, no more than five miles distant, stood the same cliffs we had kayaked past on that first night, leaving Seyðisfjörður. It heartened me to find the scenery so visually imprinted. I was itching to reach the corner where we would turn inland, short of that next headland.

Entering Seyðisfjörður, we began to count down the final ten gentle miles. As the fjord progressively narrowed, the elevation of the mountain ridges either side climbed steadily to more than three thousand feet, hemming us in. The mountains blocked the sun. For once, the water was flat and lifeless, silky calm. Yet although our mirrored kayaks slid across the placid water like birds on a glide, our progress felt impossibly slow.

I fidgeted against aches and pains I could more easily ignore when my kayak danced on a livelier surface. There was nothing to distract me or to ease my aches. Geoff paddled effortlessly beside me, his kayak reflecting red and yellow in the water. Strapped to the deck behind him was the whale vertebra he picked up weeks ago and decided to carry home. Alongside that: empty gallon plastic water bottles and a couple of gear bags.

On his front deck he carried his chart and compass. There too, a square, khaki canvas, government surplus bag, identical to the one I used. Mine carried snacks, and a small camera sealed inside a Tupperware box. On his head, Geoff wore his well-weathered canvas hat, and on his tanned face, a smile of contentment.

Patches of snow decorated the mountains, lingering, shrunken, un-melted in the shadows; the skulking remnants of the snow that lay thicker when we began our journey.

We had come full circle and yet these last few miles seemed to drag at my hull, the distance stretching as if holding us back. Time seemed to slow. As it grew darker, we could see lights burning here and there.

Suddenly, a car stopped on the road along the southern side of the fjord, and someone got out to stare. The car horn honked, the sound carrying startlingly clear across the calm fjord. We waved. Although we could not be sure from that distance, we thought we recognized the huge American sedan as one that had followed us slowly beside the fjord on our departure in June.

I heard the door slam before the car turned. Matching our own pace, it idled toward town.

As we approached the ferry quay, I could not recognize where we had started from. "This is embarrassing, Geoff. It doesn't look right," I puzzled. "We must have launched somewhere around here, but I don't remember where."

Geoff agreed. "I don't remember that big building. But I'm sure," he pointed, "that little beach, over there, is where we started." We paddled toward shore, landed together, and shook hands.

The two lads in the car greeted us and welcomed us back. They had seen us off and were excited to be the first to see us return. While we changed, they drove away to find Hjalmar, the harbormaster, who followed them back. All smiles, he shook us by the hand, congratulating us and welcoming us home.

He fetched gasoline for our stove, handing us the can before opening a door into the building on the quay. "My new warehouse," he said proudly, stepping aside for us to enter. "It's newly finished, so we haven't moved in yet. It wasn't here before you left. You may sleep in here if you like and sort your gear."

It seemed strange to carry our kayaks into the corner of this cavernous building, to spread our things on the concrete floor. Every sound we made was amplified inside the vast empty space.

It smelled of fresh cement, the same distinctive odor I remember from when I was six years old, when my parents bought the newly built house in Brighton where I grew up.

We followed Hjalmar out to his car where he took his leave, raising a cloud of road dust. Then, he braked abruptly and reversed back through the dust. "The mayor sends his best wishes," he called from his window. "I had to ask his permission for you to use the warehouse."

"Please convey our thanks!"

Our paddling trip was complete. I felt pride in our achievement yet had mixed emotions. Looking back made nine weeks seem such a brief time. Even on the map, the island looked much smaller than when we started. I recalled each stage of the journey, remembering what had happened.

Every day so far, we had studied that map, planning our next step. Tomorrow we would not. I already missed the anticipation and the routine. The motivation for our daily adventure having evaporated, I felt strangely empty.

Unrolling my sleeping mat, I realized we had no further need for the tent. With chagrin, I finally understood why Geoff had wanted to split the remaining short distance into two days, to stretch out the end, to savor the last few miles. He knew. We live for anticipation.

There was nothing else for it; I would have to plan another trip, to somewhere else.

Iceland by Kayak

GEOFF ENTERING
SEYÐISFJÖRÐUR.

AUTHOR ON FINAL DAY.

JOURNEY'S END.
PHOTO BY RUDOLPH KRAUTSCHNEIDER.

30

Seyðisfjörður Again

We could now prepare for our journey home. From Seyðisfjörður we would stay onboard Smyril, while it paused at Tórshavn, on its way directly to Scrabster. There, we would extract Geoff's dark green minivan from the harbormaster's shed. After three months idle, it might need a push down the road to bump-start it, or jumper cables, before our drive to south-east England. I was longing to see my family and friends, looking forward to getting home.

With time to spare came the opportunity to clean everything. On the quay, our kayak cockpits made great makeshift basins in which to wash our clothes. Once everything hung drying over a fence, it was time to relax.

For the time being, roast dinners, cheesecake, and other subjects of our fantasies would have to wait, but within our means we had promised ourselves a celebratory treat. We strolled into town and bought a sizeable quantity of the Icelandic yogurt, *skyr*, something we had promised ourselves if we finished the trip. We ate it all. Then, helpless to move, we fell asleep in the sun.

Iceland by Kayak

Across the fjord, a small sailing yacht lay on the dark water, forlorn against a quay with derelict warehouses. It looked so out of place that we walked around the end of the fjord, through town, to investigate.

There, in a deserted building, we located the skipper, Ruda, *Rudolf*. From Czechoslovakia, he introduced us to his crew: a vivacious Polish girl, Gosia, *Malgorzata*, and another Czech man, Pavel. They had installed themselves here, temporarily making it their makeshift home. Gosia's feet and ankles had swollen, they thought from the inactivity and cold aboard the twenty-three-foot yacht. The men were concerned for her health.

Ruda said they had sailed from the Baltic, crossing to Scotland, Shetland, Faeroe and finally Iceland. All that way in such a tiny sailboat, with a small outboard motor for alternative power.[6] I was impressed.

I liked them. They seemed kindred spirits, exploring by sea in the craft they had available. Ruda talked about playwrights, who had to be careful not to be too political or they would be in trouble with the authorities. Czechoslovakia was not the place to speak out without caution. The communist authorities had already this year cracked down on artists, prominent figures, and independent thinkers, who had protested for human rights.

Ruda had printed postcards for their trip. He corked each rolled card inside an empty bottle to drop into the water, a message in each bottle. Who knows where they might drift and who might find them? Each postcard showed an outline map of Iceland with a huge sea monster, a whale, swallowing a sailing ship. Printed were the words: *Posted on Board s/y Vela*. Vela, *whale,* was the name of their yacht. At the other corner of the card was a stylized image of a yacht with the words: SEA MAIL.

Ruda took an uncut sheet of two printed cards, stamped it with his two different rubber stamps, and then all three adventurers signed. Writing along the bottom, Rudolf asked, *"And year next?"*

Seyðisfjörður Again

"So, what about next year?" I asked when he had finished and handed it to me. "Do you have a plan?"

Ruda's dream was to build a sailboat with a steel hull to explore the arctic. "Where?" He shrugged. "Spitsbergen?" After the Arctic, he wanted to go to Antarctica. "Perhaps you would like to come?" he asked.

That reminded me of Dave, in Maidstone. Was it only a year ago when I called to him, voicing my first thoughts of Iceland? I could still hear him chuckle, thinking I was joking. So, I felt sure Ruda would build his yacht and sail those adventures. By voicing the idea, he had already begun.[7]

Geoff and I leaned over the rails of the ferry as it accelerated from the dock, exhaust fumes billowing across the deck. On the far shore floated Vela, tiny against the magnificent mountains, and appearing even more diminished from our elevated vantage point. Beside the yacht stood three figures. Whether they waved at us, or at the ferry, it was impossible to tell. We waved back, and so did other passengers on deck. Then, one of the figures raised an arm in what appeared at first to be a salute. Moments later a brilliant green flare arched up, a perfect closure to a wonderful summer.

***

I returned to Burwash Place, to teach, feeling fit and self-confident. Yet I could find no adequate way to summarize our trip. A chance encounter brought that home to me.

"Hi, Nigel, I haven't seen you in a while. What's up?"

"I've been away. I paddled a kayak around Iceland." I felt a flush of pride.

"Really? Did you meet Eskimos?"

"Inuit? No. That's Greenland."

"Oh. I thought that was Iceland. So how was it?"

"Brilliant! amazing," I trailed off. How could I even begin to describe it?

Iceland by Kayak

"Cool, yeah, we had a great holiday too. We went to Bournemouth. Stayed at a little guest house where we could walk to the beach every day. A short walk, and then the rest of the day to soak up the rays: toast the old bones! Fish and chips and a pint in the evening. Perfect! We've already decided, we'll do the same again next year." He paused, smiling in recollection.

Turning, he asked, "Yeah, so how about you? Made any plans for next summer? I recommend Bournemouth. Like I said, a cozy place to stay, super friendly people, and right there by the beach. You could take your kayak. It would be easy."

I had paddled past Bournemouth in 1975, on my solo trip along the south coast, aiming my slapping bow that day toward the Isle of Wight and the Solent. Those holiday beaches did not even get a mention in my journal that evening. "Sounds great!" I replied, untempted. Although, it did beg the question: where next? I was twenty-four years old with a head full of ideas, and little to hold me back.

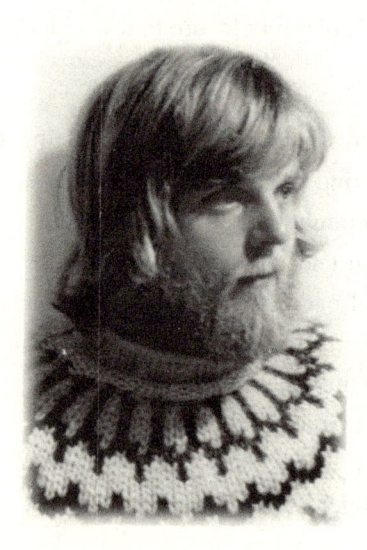

WHERE NEXT?

Epilogue

Of course, I did plan more kayaking trips. The following year I spent six weeks in Newfoundland. After that I paddled across France and explored in the Faeroe Islands and Norway. Short excursions and longer journeys alike brought me happiness.

Then, in 1981, I flew to Arctic Canada and paddled solo from Baffin Island's Iqaluit to Northern Labrador, an adventure that ended prematurely, three hundred miles from the nearest village. With frostbite in my fingers, I hitched a ride on an oil tanker to Nova Scotia. My trip stayed unfinished until 2004, when I returned with Kristin Nelson, now my wife. My book, *On Polar Tides*, (Falcon Guides) tells of Labrador, and polar bears, and all the wonders of a remote northern land.

I have since seized opportunities to explore unfamiliar territory, and revisit places close to my heart, such as Iceland. I made a career of kayaking, designing kayaks and instructing.
When Geoff returned home, he finished renovating his Sussex property and married Jeannie, the wonderful woman he met shortly before leaving for Iceland. They raised a family in the UK. Geoff's book with Bill Taylor, *The Sea Paddler's Tale,* recounts Geoff's 1970 solo trip around Britain.

Although I have reunited with Geoff seldom since 1977, it is always a joy to see him. Our conversations invariably resume where they left off. He is positive, unassuming, humble, and fun. I last saw him at Ísafjörður in northwest Iceland, in 2011, where we met with two South Africans rounding Iceland in a tandem kayak. Iceland is a wonderful place for adventure.

GREAT AUK.
MUSEUM SPECIMEN, STUFFED.

Background reading

Clark, Jim. *Angmagssalik Round Britain.*
Fergus, Charles. *Summer at Little Lava: A Season at the Edge of the World.* North Point Press of Farrar Straus and Giroux, 1998.
Hunter, Geoff. Taylor, Bill. *The Sea Paddler's Tale.* Pegasus, 2023.
Hydrographer of the Navy. *The Arctic Pilot, Volume 2, seventh edition*, 1975.
Kidson, Peter. *Iceland in a Nutshell.* Iceland Travel Books, 1974.
Scherman, Katherine. *Iceland Daughter of Fire.* Victor Gollancz Ltd. 1976.
Severin, Tim. *The Brendan Voyage.* Book Club Associates, 1978.
Storry, Terry. Bailie, Marcus. Foster, Nigel. *Raging Rivers Stormy Seas*, Oxford Illustrated Press, 1989.
Njal's Saga. Penguin Classic, 1960.

More titles by Nigel Foster

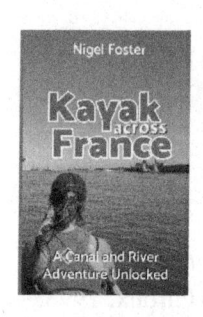

Kayak across France. (*Nigel Kayaks*). A winter dash across France, north to south, and a leisurely September jaunt, Mediterranean to Atlantic. The latter follows Foster through classic French wine country. The 17th century Canal du Midi passes historic fortifications and towns. Wine chateaux. cheeses, and baguettes, add temptations. Foster teases out background stories from Languedoc and Bordeaux on his route by canal, river, and estuary, ending with oysters and Champagne by the ocean.

Heart of Toba. (*Nigel Kayaks*). By kayak, to explore Batak Life beside the World's Largest Caldera Lake. An adventure to discover the hidden wonders that make life beside Lake Toba, and on its volcanic resurgent island Samosir, so special. (North Sumatra, Indonesia).

On Polar Tides. (*Falcon*). Paddling and surviving the wilderness of Northern Labrador. and Ungava Bay; realm of polar bears, fierce mountain squalls and extreme tides. Woven from two kayak journeys are tales of early exploration, plane wrecks, and meetings with polar bears and Inuit seal hunters. A World War II German weather station lurks on a remote island, installed by a submarine crew where green auroras twist in the night sky, and fog drifts around stately icebergs.

Encounters from a Kayak, (*Falcon*). Thirty-nine stories from Foster's diverse experiences around the world: tales that reveal all that makes kayaking special. Arranged into four sections, the stories tell of Creatures, People, Special Places, and Flotsam and Jetsam. Color images throughout.

Paddling Southern Florida. (*FalconGuides*). An updated guidebook to more than fifty paddling trips in Southern Florida. Includes road access to launch and finish, detailed route guidance with points of interest, suggestions for accommodation, dining, and rental options. Sidebars offer handy details about Florida wildlife, weather, and history. Priceless if you are to maximize your trip to Florida.

The Art of Kayaking. (*Falcon*). All you need to know about paddling. From necessary skills to essential gear, this is the distilled work of Foster's innovative instructing, expeditions, and kayak design experience. Find sound advice for beginners, alternative techniques, subtleties for finesse, and fine control for the advanced. Color photo action sequences, maps, and diagrams, make this an easy-to-follow book for all levels, beginner to expert. On calm water or rough, there is something for everyone.

More at
www.nigelkayaks.com

End Notes

[1] **Chapter 5, page 48.**

Smyril IV was named *MV Morten Mol*s when first built, in 1969, for the Danish shipping company Molslinjen. Strandfaraskip Landsins bought the ship, in 1975, to run from the Faroe Islands.

By the mid-1990s, the ferry showed signs of fatigue from working in the harsh, north Atlantic. Struggling by 2003 to face the strong Faeroe Islands currents, she ceased operation in October 2005. *Smyril V* replaced her.

[2] **Chapter 17, page 183.**

In a series of small coincidences, Tim Severin, and his crew aboard the leather curragh *Brendan*, seemed to skip ahead of me through the summers of 1976, 1977, and 1978.

Brendan began its voyage north from Ireland in 1976, stopping at Port Maddy in the Outer Hebrides, on route to the Faeroe Islands and Iceland. Dave and I paddled from Skye to Lock Maddy later that summer, where we ran into Colin Mortlock and his kayaking friends, and shared a pot of tea.

In 1977, when Geoff and I paddled around Iceland, fishermen told us how *Brendan* overwintered there, but had since left for North America. When *Brendan* completed the voyage, landing at Musgrave Harbour, Newfoundland, on June 26th, Tim Severin proved to the world that Irish skin boats could make the voyage from Ireland to North America. They may have made the journey long before the Vikings did.

Iceland by Kayak

In 1978, while kayaking in Newfoundland and reaching Musgrave Harbour, I met fishermen who spoke of *crazy Irishmen* arriving a year ago in a skin boat. They shook their heads in wonderment at how two *crazy Englishmen* were now exploring Newfoundland by kayak.

A few years later when kayaking in the Faroe Islands, I was honored to meet the artist Tróndur Patursson, at his Kirkjubøur home. It was there he had joined the Brendan voyage to replace one of the original crew. Brendan was not the only adventure he had with Tim Severin, so he regaled me with stories of some of their later seagoing exploits.

[3] **Chapter 24, page 274.**

What Geoff regretted tasting puzzled me, as did the unidentifiable chunks of flesh hanging beside the fish in the drying shed. The mystery was solved when I returned to Iceland, eight years later. Kayaking with a group of teenagers in the Westfjords, we camped in a small fjord north of Gjögur. There we met two men who introduced me to *hakarl*, rotten shark meat.

They were one of the only families in Iceland to produce this delicacy. They bury a Greenland shark in the shingle beach for six months. When it has sufficiently decayed, they dig it up, cut it into chunks and hang it, for up to six months, to air. This rids the rotted meat of some of its putrid ammonia smell.

Cut into small cubes, it fetches a premium price in high class restaurants in Reykjavík, where diners take it as an hors d'oeuvre, with a shot of Brennivín, Icelandic schnapps. For more, read the story in my book, *Encounters from a Kayak*, (Falcon Guides).

[4] **Chapter 25, page 281.**

Iceland remained under Norwegian sovereignty until the Kalmer Union of 1380 between Sweden, Denmark, and Norway which included Iceland, the Faeroe Islands, Shetland, and Orkney.

When the Kalmer Union split up, Denmark was left in control of Norway and Iceland. When Norway became independent, Iceland remained a Danish colony until 1944.

[5] **Chapter 28, page 326.**

Denarius derives from dēnī, *ten*. One silver denarius coin, introduced in Rome in 211 BC, was worth ten *asses*. The *as* was a weight-based coin cast from one Roman pound of copper, about eleven-and-a-half ounces, (325g). The old British penny took its symbol *d* from denarius.

Britain has used the pound, as a unit of currency, for more than 1,200 years. From the Latin, libra pondo, loosely meaning *scale-pound,* its value was originally that of a pound of silver. The symbol L or £, stands for the Latin *libra:* scales/balance.

When Britain decimalized its currency in 1971, combining a Roman legacy with a modern European standard, it kept the pound as a unit, but introduced the new penny, valued at one hundred pence per pound, denoted by an unambiguous *p*.

[6] **Chapter 30, page 348.**

In 2017, Malgorzata Krautschneider, *Gosia*, living in USA, published her story of their 1977 sailing voyage to Iceland. From her book titled: *The Vela*, I learned of the storm they narrowly survived on their return journey.

[7] **Chapter 30, page 349**.

Ruda finished his steel-hulled yacht and sailed away for a lifetime afloat. Trapped on the Falkland Islands when Argentina took them in 1982, it seemed unlikely he would get permission to leave. Happily, I heard from the British authorities when the Argentinians finally let him sail away, and I relayed the good news to his anxious family.

That was just one of Ruda's escapades, adventures which made him a sailing celebrity in the Czech Republic. He is a man with worthy tales. He has written books and promotes sailing for young people. His book, *Around the World for the feather of a Penguin*, (Rudolf Krautschneider) is a gem, published in English in 1999.

Made in the USA
Middletown, DE
03 March 2024